T0332916

Darwinian Sociocultural Evolution

Solutions to Dilemmas in Cultural and Social Theory

Social scientists can learn a lot from evolutionary biology – from systematics and principles of evolutionary ecology to theories of social interaction including competition, conflict and cooperation, as well as niche construction, eco-evo-devo, complexity, and the role of the individual in evolutionary processes. Darwinian sociocultural evolutionary theory applies the logic of Darwinism to social-learning-based cultural and social change. With a multidisciplinary approach for graduate biologists, philosophers, sociologists, anthropologists, social psychologists, archaeologists, linguists, economists, political scientists and science and technology specialists, the author presents this model of evolution drawing on a number of sophisticated aspects of biological evolutionary theory. The approach brings together a broad and inclusive theoretical framework for understanding in the social sciences which addresses many of the dilemmas at their forefront – the relationship between history and necessity, conflict and cooperation, the ideal and the material and the problems of agency, subjectivity and the nature of social structure.

MARION BLUTE is Emeritus Professor of Sociology at the University of Toronto at Mississauga where she has taught classical and contemporary theory and gene–culture coevolution to undergraduates. She also has taught contemporary theory in the university-wide graduate sociology programme. She has published in a wide variety of life and social science journals on evolutionary topics and has related interests in the philosophy and sociology of science. She is a member of the Editorial Advisory Board of *Biological Theory* and of the Editorial Board of *Spontaneous Generations: A Journal for the History and Philosophy of Science.*

There is a grandeur in this view of life, with its several powers, having been originally breathed into a few forms or into one; and that, whilst this planet has gone cycling on according to the fixed law of gravity, from so simple a beginning endless forms most beautiful and most wonderful have been, and are being, evolved. (Charles Darwin, concluding words from the first edition of *The Origin of Species*, 1859.)

Biologists ... appreciate the hugeness of the problem that Darwin faced and solved. They are therefore more likely than social scientists to feel optimistic about the chances of a comparable intellectual feat in the study of cultural evolution. (John Maynard Smith and N. Warren, *Evolution* **36**:3, 620, 1982.)

That there are just two major forms of behavioural evolution, occurring through genetic and cultural transmission respectively, must rank among the most exciting and fundamental discoveries ... achieved over the last century and a half. (Andrew Whiten, *Behavioral and Brain Sciences* **24**:2, 359–60, 2001.)

Darwinian Sociocultural Evolution

Solutions to Dilemmas in Cultural and Social Theory

MARION BLUTE

CAMBRIDGE
UNIVERSITY PRESS

Shaftesbury Road, Cambridge CB2 8EA, United Kingdom

One Liberty Plaza, 20th Floor, New York, NY 10006, USA

477 Williamstown Road, Port Melbourne, VIC 3207, Australia

314–321, 3rd Floor, Plot 3, Splendor Forum, Jasola District Centre, New Delhi – 110025, India

103 Penang Road, #05–06/07, Visioncrest Commercial, Singapore 238467

Cambridge University Press is part of Cambridge University Press & Assessment,
a department of the University of Cambridge.

We share the University's mission to contribute to society through the pursuit of
education, learning and research at the highest international levels of excellence.

www.cambridge.org
Information on this title: www.cambridge.org/9780521768931

First published 2010

A catalogue record for this publication is available from the British Library

ISBN   978-0-521-76893-1   Hardback
ISBN   978-0-521-74595-6   Paperback

Contents

Preface: a postmodern metanarrative

This book has three goals. This first is to acquaint colleagues in the social sciences and their students who may not be well informed, but who want to be become more so, about work currently being done in a variety of social science disciplines on Darwinian theories of socio-cultural evolution. These are the readers who, when they hear "cultural evolution", think "sociobiology", or even of the kinds of stage theories of "progress" common in the nineteenth century and the first half of the twentieth century. Conversely, life scientists may be interested in the extent to which Darwinian evolutionary concepts and theories are applicable to purely sociocultural phenomena.

The second is addressed to colleagues in either the social or the life sciences and their students who know better and are currently working in this area. I would like them to become aware, to the extent to which they currently may not be, that their labours in their own discipline are supported by, and converging with, those of others, working in virtually every other social science discipline. Together they are building the foundations of a broad and inclusive theoretical framework for understanding in the social sciences. I would also like to encourage an enriched and updated understanding of evolutionary processes beyond the heretofore admittedly useful formulas such as those of replication, variation, interaction and selection many of us have been working within. Today, evolutionary biology has a richer conceptual apparatus to offer sociocultural evolutionists that we need to take advantage of. These include developments in systematics, the emergence of general principles of evolutionary ecology, the role of conflict in a cooperative context and vice versa, niche construction, and an awareness that a truly synthetic theory must embrace not only transmission and selection but also development and ecology if it aspires to explain the evolution of complexity. This is not to say that many problems do not remain outstanding – those pertaining to the role of the individual in evolutionary processes for example.

The third is to use the second to inform the first. I would like to show how an enriched understanding of the sociocultural evolutionary process can address many of the dilemmas that are at the forefront of the social sciences today. What is the role of history and contemporary social forces in understanding human affairs? What conditions are likely to result in social conflict or cooperation or both? What about memes? What about rational choice versus reinforcement? Is everything, something, or nothing socially constructed? Where does complexity come from? What is social structure anyway and how does it relate to culture? Is there a role for human agency in a culturally programmed and socially structured world?

Nothing so characterizes the postmodern state of the social sciences as the belief that metanarratives (in the language of the humanities) or general theories (in the language of the sciences) are impossible. At first blush therefore, the title of this preface is a contradiction in terms. Sociocultural evolutionism is most definitely a metanarrative, a general theory. At the same time, however, it is a postmodern one. It does not seek to outcompete, let alone actively destroy, other theories, paradigms, theoretical orientation or schools of thought dealing with human culture and social organization. Rather, it is a framework which respectfully acknowledges fundamental insights of the most diverse sorts and sources about our common subject matter. Evolution acknowledges the significance of both the ideal and the material, change and stability, history and necessity, cooperation and conflict, reason and reinforcement, the subjective and the objective, and hopefully ultimately both the biological and the sociocultural. Evolution is indeed then a *postmodern* metanarrative, a framework within which what we all have to say makes sense. This book is about that metanarrative, that general theory.

Acknowledgments

The author would like to express gratitude for the teachers whose dedication to cultural and social theory inspired – Daniel Rossides, Jos Lennards, Lewis Feuer, Anatol Rapoport, Bonnie Erickson, Richard Elinson and Wsevold Isajiw; to best friend Gail Greer who has been an intellectual companion on every step of this journey; to family, both natal and current, without whom it would not have been worth it; to students and colleagues in the Sociology departments of the University of Western Ontario and the University of Toronto who patiently tolerated commitment to a "big idea"; and to the literally hundreds of colleagues, evolutionists and otherwise, in many disciplines and countries, alongside whom I have grappled with ideas in person and in print over the years. Specifically I would like to thank Alison Dias who prepared the figures and tables and Sam Clark, Sigrid Glenn, Nikolaus Ritt, Tang Shiping and Jonathan Stone who kindly read and commented on various chapters, the responsibility for which of course remains my own.

1

Introduction

In 1982 I had the pleasure of riding in a train across part of southern Ontario in Canada with the late Bill Hamilton who many think introduced the greatest innovation in the theory of evolution since Darwin – the theory which came to be known as "kin selection" or "inclusive fitness" (1964 a,b). Hamilton had pointed out that selection would act on (and hence calculations of fitness should take into account) not only the effect of our genes on our own behaviour, but also their effect on relatives, because the latter, to varying degrees depending upon the relationship, are carriers of the same genes identical by descent. For example, a gene which influenced one to assure the survival of a little more than two full siblings at the cost of one's own life would be favoured by selection because, on average, it would be transmitted through a relative rather than personally. His insight, model and initially suggested applications went on to give rise to a vast lineage of research on cooperation among relatives in nature. It was this work that stimulated Edward O. Wilson to write his *Sociobiology: The New Synthesis* (1975) which had caused such a stir among social scientists while I was a graduate student.

We were leaving a conference in Kingston Ontario and ended up on the same train. I was getting off at my home, Toronto, while he was going on further. I did not want to bother him and took a different seat but he came over and joined me. We discussed a variety of things in a few hours. He kindly relieved my embarrassment at having presented a paper at the conference which reinvented something that, unbeknownst to me at the time, part of which he had already published. It was about some implications of asymmetry in sex chromosome inheritance. Instead of adding to my embarrassment he complemented me on having got the numbers right. He also wondered why, in the human species, it is females who tend to ornament themselves, while in most other species it is males. I suggested the former impression might be a result of sampling error and not representative of the diversity of human societies and cultures across the five to seven million years of history since we diverged from our common ancestor with chimpanzees, and provided a few examples. We also discussed his theory that antagonistic coevolution with parasites is responsible for the maintenance of sex – sexual species may manage to resist the onslaught of their much more rapidly evolving parasites by recombining their genes in every generation, making parasite populations, in effect, start their "pursuit" all over again. In making that argument he had noted that our first line of defence is the adaptive immune system which functions in many respects as a mini-evolutionary process taking place within the organism. I explained that the individual learning process by reinforcement and punishment is also such a miniature evolutionary process which takes place within the organism.

But most of all, we discussed the landscape we were travelling through – particularly the fact that it was at least as much marked by culture as by biology. We noted the land strewn with human artifacts – the roads, tracks, overhead wires, fences, farm houses, villages and towns, domesticated species, and the cleared (and not uncommonly again abandoned) fields being replenished with a new ecosystem of mixed native and once cultivated species gone wild. I was subsequently very surprised and disappointed then, as the three volumes of his collected works accompanied by reminiscences of their genesis and publication appeared after his tragically early death, to learn of his shockingly out-dated and dangerous eugenicist views. Parents should be free to practise selective infanticide; those who want to keep a "vegetable baby" alive should be required to pay for it (2001 V 2:xiviii); caesarean birth rates put us at risk of evolving a species incapable of giving birth naturally and so on he argued. How could one with so kindly and gentle a nature, in both my own experience and the accounts of others, be so wrong in ethical matters as to endorse views that in the past have contributed to discrimination, forced sterilization, murder and genocide? With respect to science rather than ethics, how could a scientist who had been so brilliantly original in his papers and who had chatted so freely with me about culture, ultimately have been so obtuse on the topic of psychology and the social sciences? As one reviewer of the second volume (Hamilton 2001) put that point:

Hamilton's line connecting the behaviour of animals to that of
humans was short and straight, the evolution of the human brain
having done little to warp or extend it. His biologisation of our species
accordingly pervades the (first) two volumes, but is more prominent in
the second (Barry 2003).

It made me wonder for a time if the critics were right, if there was
something intrinsic to evolutionary theory that fostered such views.
After the initial shock wore off, I remembered that for every prominent
evolutionist who expressed such eugenicist views, there were more
who denounced them (e.g. Gould 1981, 1996; Lewontin, Rose and
Kamin 1984). For every evolutionary biologist who thought that genes
count for everything and culture for nothing, there were similarly
renowned ones who made a point of acknowledging the potential
significance of cultural evolution (Dawkins 1976:191–201, 322–31;
Maynard Smith and Warren 1982; Williams 1992:15–16, 18–19;
Maynard Smith and Szathmary 1995:309; Futuyma 1998:4) and some
had made it a focus of their work (Cavalli-Sforza and Feldman
1981; Ehrlich 2000). For these kinds of reasons, it is important that
I make clear right at the beginning what this book is not, as well as
what it is, about.

First it is not about the application of gene-based biological evolutionary theory to human behaviour (sociobiology, human behavioural
ecology, evolutionary psychology, etc.) which is not to deny its utility.
Shorn of racist and eugenicist views, as with a few rare exceptions
it currently is (albeit sexism is another story), evolutionary biology
applied directly to humans has yielded a great many interesting
insights about human nature, the history of human populations and
even their health.

1.2   DEVELOPMENTALISM

Secondly, it is not about the developmental stage theories of historical
"progress", what Karl Popper (1957) called "historicism", and that have
been called evolutionism for most of the history of the social sciences.
In those theories, societies, cultures or particular institutions are proposed to necessarily pass through (some) particular sequence of stages.
For example:

Societies or cultures were held to develop from despotism through
monarchism to republicanism (Montesquieu), from the theological
through the metaphysical to the scientific (Comte), from status to
contract (Maine), from the primitive through the feudal and capitalism

to the socialist (Marx), from savagery through barbarism to civilization (Morgan), from gemeinschaft to gesellschaft (Toennies), from the ideational through the idealistic to the sensate (Sorokin), from folk through feudal to industrial (Redfield), from mechanical to organic solidarity (Durkheim), etc. The family was proposed to develop from sexual promiscuity through the matrilineal family through the patrilineal family to the conjugal family; the economy from gathering, through hunting, through herding through agriculture to industry; technology from wood, through stone, through bronze to iron; legal systems from communal to private property or from status to contract; religion from magic through animism through totemism to belief in a personal deity; and for the polity there were many sequences but all ended in democracy. (Blute 1979:47)

Sanderson's excellent recent history of evolutionism in sociology and anthropology (2007) sees slightly modified versions of such stage theories of history as the essence of evolutionism in the social sciences. He adopts Wright's (1983) criteria that evolutionary theories must have three features.

1. It must propose a typology of social forms with potential directionality.
2. It must order these social forms in the way it does on the assumption that the probability of remaining at the same stage in the typology is greater than the probability of regressing.
3. It must assert a probability of transition from one stage of the typology to another. (2007:5)

In evolutionary biology, passing through a characteristic sequence of stages is something that *individual* members of evolving populations rather than populations, species or higher taxa themselves do. Such theories are more akin to developmental than to evolutionary biology and are more aptly characterized as such. This is not to claim that so many distinguished social scientists were wholly wrong. For example, hunting and gathering wild food *was* the primordial mode of human subsistence. Horticulture (small-scale agriculture with human labour using hand tools), intensive agriculture (farming with energy from animals yoked to plows), and industrialism (in which human and animal labour is largely replaced by machines using fossil fuels) *did* branch off successively later (Blute 2008b). However, the emergence of such novel modes of subsistence accompanied by new forms of social organization are more akin to what biological evolutionists historically called "grades" and more recently have called "major transitions" (Maynard Smith and Szathmary 1995) – transitions such as the emergence of the more complex eukaryotic from simpler prokaryotic cells, or of multicellularity from unicells, than they are to any necessary sequence of stages universal to some class of units. Such increases in complexity in some lineages (or even progress if one's values lead one to label them as such) are part of the story of evolution and that should not be forgotten, as Arthur (2006) has recently emphasized for biology. However, while the logic of the evolution of such increases in complexity is discussed here in Chapter 8, they are only one part of the larger picture of innovation and recombination, differential proliferation under selection, and diversification of species, which characterize the branching tree of an evolutionary process.

These traditional theories were commonly developmental in a second sense in that the process of change itself was commonly understood in terms that a biologist would recognize as developmental rather than evolutionary. Multicellular development and evolution are deceptively similar – both involve growth (the proliferation of cells in development, of individuals in evolution), a branching process (differentiation in development, diversification including speciation in evolution), and interactions among the "branches" (normally cooperation based on a division of labour in development but any or all of competition, conflict and cooperation in evolution).

A classic example of such developmental theories of social change was the work of the English polymath, Herbert Spencer, perhaps the most misunderstood theorist in the history of the social sciences. Stereotypes would have it that Spencer took Darwin's theory of evolution and tried to apply it to society. In fact, rather than being inspired by Darwin, Spencer was a believer in free enterprise from the beginning as was made clear in his first book, *The Proper Sphere of Government* (1842) which expressed his faith in nineteenth-century liberalism, the common intellectual coin of the time in England. Subsequently, his famous essays on "the development hypothesis" (which could profitably be read by evolution-deniers today) and on "Progress: Its Law and Cause" (making clear the developmental nature of his theory of change) were published seven (in 1852) and two (in 1857) years respectively before the first edition of *On the Origin of Species* in 1859. Reviewing a textbook of physiology drew his attention to the work of the German embryologist Karl Ernst Von Baer on differentiation in development. Rather than Darwin, Von Baer was the immediate inspiration for Spencer's "theory of everything" that progress consists of transitions from the homogeneous to the heterogeneous accompanied by integration – as he makes clear in his autobiography (1904, II, 8–13 & 165–170) – and was the source of his expectation that free competition would result in cooperation rather than conflict.

> From the lowest living forms upward, the degree of development is marked by the degree in which the several parts constitute a cooperative assemblage (1862:276).

> Nothing like a high type of social life is possible without a type of human character in which the promptings of egoism are duly restrained by regard for others (1873:198).

Spencer believed that governments should not provide poor relief for example, not because the competition required for progress demanded that the unfit go to the wall so to speak, but because he thought that government interference would frustrate the development of private charities. Towards the end of his life Spencer was depressed that people were not living up to his expectations that they would create the cooperative "social state" which he had envisaged and became something of a misanthrope. Some (e.g. Peel 1971:137) have emphasized an origin in social rather than biological science for Spencer's emphasis on specialization and cooperation – Adam Smith by way of Milne-Edwards on "the biological division of labour" which Spencer also mentions. There can be no doubt, however, that while mentioning this, Spencer's autobiographical account emphasizes the embryological influence. Again, be what it may, Ghiselin got to the heart of the matter. Spencer "assumed that individuals act in the ultimate interest of society" (1974a:224). Not exactly assumed, however – rather cooperation emerges as a consequence of crowding, and among people for example, characterizes industrial society.

Later, the conception of change of the theorist most responsible for the institutionalization of sociology in the nineteenth century, Emile Durkheim, was similarly developmental, and indeed indebted to, Herbert Spencer (Corning 1982). In *The Division of Labour in Society* (1893) Durkheim conceptualized society as being, like an organism, an entity which passes through stages while it grows (increasing in density and frequency of interactions) and develops. It develops in the sense that it becomes (a) increasingly internally more heterogeneous, for example in its occupational structure, and (b) increasingly internally more functionally interdependent – all changes which Durkheim called the transition from "mechanical" to "organic" solidarity. He offered as evidence for such a transition the increasing quantity and importance of contract law regulating relations between individuals and groups as opposed to criminal law. He offered as an explanation for the transition – need. Growth makes specialization and cooperation necessary and they appear because they are needed. Durkheim's *Elementary Forms of the Religious Life* (1912) is in the same vein – religion appears in the historical development of human societies universally because it is needed for social solidarity.

Later developmentalism was the source of inspiration for the general systems theory of Bertalanffy (1968) whose earlier work was on developmental biology and for his followers in sociology (Buckley 1967). It was also the version of functionalism finally settled on by Talcott Parsons (1966; albeit like so many others, he called it evolutionism). According to Parsons (1973:72) his functional theory of change was "most closely analogous to the process of growth in the organism", a "process of structural differentiation and the concomitant development of patterns and mechanisms which integrate the differentiated parts". Indeed, the developmental influence continues to linger at least terminologically, if not otherwise, in the use of terms like "development" (found everywhere) and "morphogenesis" (e.g. Archer 1995) derived from embryology for cultural and social change.

1.3   DARWINIAN SOCIOCULTURAL EVOLUTION

If it is not about sociobiology or about developmental-stage theories of historical progress, what is this book about? Contemporary Darwinian rather than developmentally inspired theories of change in the social sciences come in three broad forms – the gene-based biological (sociobiology, human behavioural ecology and evolutionary psychology), the social learning or meme-based sociocultural, and dual inheritance or gene-culture coevolutionary theory. This book is about the second – about sociocultural evolution. Again, this is not to deny the utility or importance of the others (in my own career for example, I have published in all three areas). However, I am most excited about the possibility of a unification of the social sciences themselves within a broadly synthetic sociocultural evolutionary framework. Ultimately, of course, both life and social scientists will also have to face up to the interaction between evolution in the two realms. While some progress has been made on the topics of dual inheritance and coevolution (e.g. Boyd and Richerson 1985; Durham 1991; Feldman and Laland 1996; Richerson and Boyd 2005; Blute 2006a), there is still a long way to go. While the relationship or interaction between the two is not the subject of this book either, it will be returned to in the final chapter on the future of the social sciences.

Sociocultural evolutionary theory in the second sense, which is the subject of this book, the social learning or meme-based variety, has been bubbling up from the bottom in recent decades. It is being developed and applied in virtually every social science discipline. These include the study of languages, technology both prehistoric (archaeology) and historic, science, economic organizations and institutions and memetics as well as, to some degree, in the most general social science disciplines including anthropology, sociology and even history. We will not so much be considering the story of sociocultural evolution, which, after all, is the story of all of human history, but the process. By now, a rather large literature has accumulated and there is fairly wide agreement on the most basic elements of the process – commonly described after Hull (1988), for example, as replication, variation, interaction and selection. Replication is commonly thought to take place socioculturally by any mechanism of social learning (but see Chapter 5). Typically, it involves the transmission of ideas, (or "memes" suggesting memory and genes in Dawkins 1976 terminology) – "information such as knowledge, beliefs, and values that is inherited through social learning and expressed in behavior and artifacts" (Mesoudi, Whiten and Laland 2004, paraphrasing Boyd and Richerson 1985). Sociocultural evolution, at least when social learning takes place by linguistic instruction rather than simply by observation (Blute 2001a), includes an equivalent of the distinction between genotypes (digitally encoded information such as the genes influencing height) and phenotypes (observable characteristics such as height itself) in biology. Variations, including new combinations, are introduced from time to time.

These socially learned "iss" and "oughts" informing and directing behaviour have been given many terms in various social sciences through their history. Sociologists refer to "norms and values". Anthropologists once liked to talk about "folkways and mores" but today more often say "traditions" (albeit some students of animal behaviour like to call animal cultures "traditions" to distinguish them from human culture). Linguists speak of "rules" or "competencies"; institutional, including evolutionary economists and organization theorists of "conventions", "habits", "routines", and "competencies". Archaeological speak is quite varied – "techniques", "design elements", "traits" and "traditions", for example, are fairly common. In science studies they speak of concepts, theories and methods but also of more inclusive entities such as "research programmes" and "paradigms". In any event according to evolutionists, such socially learned information and instructions, commonly embodied in social roles, statuses or identities, interact with each other, with other inherited resources such as wealth, income, power and status and with the environment yielding the visible behaviour and artifacts characteristic of some particular social identity. As a result of this, some social identities prove more viable and more successful at recruitment than do others, and hence their beliefs and values become relatively more common, i.e. selection and evolution in the form of statistical changes in a population take place. Molecular biologists out-compete cell biologists; specialists doctors out-compete general practitioners; born-again Christians and Islamists out-compete more traditional denominations of their religions and so on. As with viruses in biology, in an era of mass communications particularly, information more or less disembodied from particular social identities can also spread horizontally and evolve as well, a fact often emphasized by memeticists to many of whom a meme is simply "an idea that spreads". "Information wants to be free" as Stewart Brand so famously declared and Web enthusiasts constantly remind us.

Some differences of opinion remain, of course – about the choice of terms as above and about the mechanisms involved in social learning or imitation, for example. Some view sociocultural evolution as analogous to the biological, while others prefer to think in terms of evolutionary epistemology (Campbell 1974), universal Darwinism (Plotkin 1994; Cziko 1995; Dennett 1995) or multi-process selection theory. This is the theory that all knowledge-acquiring and utilizing processes are selection processes and includes individual learning by reinforcement and punishment and certain aspects of the adaptive immune response (Hull, Langman and Glenn 2001) as well as biological and sociocultural evolution as special cases. When they first come to this kind of evolutionary theory in the social sciences, one of the first thoughts that often occurs to people is that cultural evolution is "Lamarckian" rather than "Darwinian". Superficially, the Lamarckian view seems plausible – individuals can learn things, by trial and error for example, and these "acquired adaptations" can be inherited as others learn socially from them.

The intricacies of the Darwinian versus Lamarckian argument in cultural evolution may be pursued in the literature by those interested (e.g. Murmann *et al.* 2004; Hodgson and Knudsen 2006; Kronfeldner

2006 and references therein) and is discussed here in Chapter 2. However, the bottom line I believe is three empirical issues. First, one may be uncomfortable with the notion that human innovation is "random" (meaning only "blind" i.e non-prescient, as Campbell 1965 put it). Even so, the fact is that there is no evidence in any area of human cultural activity that innovation is adaptive in the sense that it is *statistically* biased in the direction which would be required for it to spread further in the circumstances. In fact, the evidence is quite to the contrary – whether considering scientists publishing papers that will be cited, inventors obtaining patents that will be utilized, entrepreneurs founding new businesses that will succeed, or manufacturers introducing new products that will be successfully marketed (Blute 1979). Most cultural innovations, like most biological mutations, fail. Even cultural innovations that do succeed often do so in a niche quite different from that which their originator anticipated. Secondly, while acquired characteristics may be inherited culturally (as they may indeed be inherited biologically; Jablonka and Lamb 2006), they are not necessarily. A good example is religious beliefs. We may inherit religious beliefs from our parents as part of a social identity as a church member and then abandon or change these as we grow up. Not uncommonly, however, we send our children for religious instruction anyway, i.e. we do not pass on what we have since acquired but what we originally inherited. Thirdly, in the context of multi-process selection theory, when an innovation is adaptive on one level, for example rewarding to an individual, that does not mean it will be successful on a more inclusive level, for example socioculturally. A scientist may learn something in his or her own research but have difficulty persuading colleagues. Juan Delius once pointed out in conversation that masturbation and nose picking may be rewarding but they are not successful social norms!

Lamarck is remembered as having introduced the first theory of biological evolution that at least tried to explain adaptations. Environments create a need – trees grow taller; giraffes adapt by stretching their necks; and their offspring, as a consequence, inherit longer necks on this view. In suggesting there might be a naturalistic explanation for adaptations and that, in striving for complexity, it is the need to adapt to the environment that drives organisms off the medieval "great chain of being" (Lovejoy 1936) creating the tree of life, one might argue that Lamarck raised the question that made Darwin's achievement possible. Most nineteenth- and early twentieth-century biologists, including Darwin, accepted such scenarios, although

Darwin thought that his natural selection principle was more important. Unfortunately adaptations were exactly what Lamarck's theory
did not successfully explain. In the face of environmentally induced
variation and change, why would individual organisms necessarily
succeed in achieving adaptation? Why would they not just fail and
die instead? The environment induces mutilations at least as often as
it induces adaptations, and even with some inheritance of acquired
characteristics as is now known to be the case, to think that adaptations as opposed to maladaptations would be preferentially inherited is
to engage in magical thinking – "skyhooks" rather than "cranes" in
Daniel Dennett's (1995) terminology. Eventually the Darwinian view
prevailed. There is inheritance but with variation (mutation and recombination); some of this inherited variation affects viability and fecundity; and the resulting statistical shifts in characteristics between
generations are therefore attributable not to the inheritance of
acquired adaptations, but to what both Darwin and Wallace came to
call "natural selection". Increasingly, such a view is coming to prevail
in the study of cultural evolution as well. All in all, as a colleague who
began defending the Lamarckian but ended up defending the
Darwinian view of cultural evolution and who shall remain nameless
put it to me – "It is just something one has to get over isn't it?".

1.4   TWO GREAT PRINCIPLES

Donald Campbell (of "evolutionary epistemology" fame) once told me
with a twinkle in his eye that he did not know of any Darwinian sociocultural evolutionist who had not begun first by becoming familiar
with the theory of biological evolution. The twinkle of course was
because he was using the case of sociocultural evolution itself as an
example of the reality of evolution, of descent with modification, in the
realm of culture, the culture of science in this case. Darwinian sociocultural evolutionism has indeed descended from Darwinian biological
evolutionism. So while this book is about sociocultural rather than
biological evolution, that does not mean that it does not contain a
very considerable amount of biology. On the contrary. Whether one
prefers to characterize Darwinian sociocultural evolution as an analogy with biological evolution or both as tokens of the same general
type, selection processes, there are many important basic evolutionary
principles developed in biology with which the reader will become
familiar in what follows in the next eight chapters. Here we will
begin with some basics.

The most fundamental question that evolutionary theory asks and seeks to answer is, "What explains the existing array and proportions of alternatives in a population?". The answer broadly speaking is constraints, chance, history and necessity in the form of selection (Blute 1997). The laws of physics and chemistry impose universal constraints on all life and culture – it might be adaptive for a bird or a plane to fly at the speed of light but none ever will. Other kinds of possible constraints are more controversial. Some would have it that transmission rules (in biology, the laws of genetics, in culture, social learning mechanisms) constrain what evolves. However, others argue that the rules themselves (e.g. genetical systems, cultural norms about how and from whom to learn) evolve, and therefore are to be explained rather than being explanatory. Similarly, every entity that evolves also develops, i.e. has a life cycle. Some would argue that there are physiological/ developmental principles such as allometry (the tendency for changes, particularly in the sizes of parts, to take place in a correlated fashion) which constrain what evolves. However, others argue that development including allometric relationships evolve, and hence it too is to be explained rather than being explanatory (e.g. Frankino *et al.* 2007). There seems little doubt that functional interactions (whether described in genetical or physiological/developmental terms or necessarily in both according to Futuyma 1998:672), interactions such as a strongly positive correlation between large size and dense bones or a strongly negative correlation between flight as a mode of locomotion and dense bones can have effects such as a tendency towards stasis in one lineage, parallel changes in more than one lineage, and less than ideal adaptation. Some, however, would prefer to view these effects as owing to selection rather than constraints, selection stemming not solely from the ecological environment to be sure, but also from social interactions within the genome/organism itself – own condition-dependent selection.

When the relationship among various selection processes concretely is involved, things become considerably more complex. Does biological evolution constrain what can be acquired culturally or do genes and culture coevolve? Are biologists and social scientists – on the same side for a change – right that biological and sociocultural evolution "programme" what individuals can learn, or do unique individual learning experiences alter circumstances sufficiently that they become a causal factor in what evolves biologically or socioculturally? As noted previously, I do not intend to pursue these kinds of questions about concrete interactions among processes until the last chapter, except to note here that everyone at least agrees that there are *some* constraints on all selection processes, particularly the laws of physics and chemistry.

Chance also plays an explanatory role in all evolutionary processes. As the synthetic theorists of evolution beginning in the 1930s always emphasized, evolution is an "opportunistic" process. In addition to being physio-chemically possible, which, say frictionless wheels are not, something has to arise in the first place before it can be selected. Wheels are a very efficient mechanism of transportation yet evolved only in very restricted forms biologically (the rotary base of prokaryotic flagella which are the organs of motility in these simple cells and the structure of tumbleweed for example.) We can well imagine, however, that many hooved animals grazing on grasslands might well have benefited from having wheels for hind feet – perhaps pushing themselves along with their front legs in a fashion similar to the way poles are used by rafters. However, it just never happened. Chance in the more technical sense plays a role in the form of sampling error or "drift" as it is called. Particularly in populations that are divided up spatially into a large number of small, but still somewhat connected local populations, or in populations that have temporally undergone a "bottleneck", sampling error can take place. Sampling error mainly affects rare variants. Alternatives that would otherwise be adaptive can sometimes be lost or ones that would otherwise not be adaptive can sometimes be maintained in such circumstances. At the same time, no population is infinite in size so chance, to some degree, is omnipresent. While all evolutionists pay at least lip service to the importance of constraints and chance, the majority emphasize what Darwin called his "two great principles" – the "unity of types" (or history, i.e. the eighteenth-century "natural system" of classification interpreted genealogically) and the "conditions of existence" (or necessity, as Monad 1971 called it, i.e. selection). These were the twin pillars of Darwin's theory that he summarized as "descent with modification".

Duthie (2004) explained in the simplest possible terms that an evolutionary science of culture or memetics deals with these two "trademark" questions of origins and selection – where did something come from and what were the selection pressures that favoured its spread? He illustrated this with two simple technological cases – the fork and the paperclip. You might be surprised, as I was, to learn that forks diverged from knives rather than spoons. The practice developed of eating with two knives, one to cut and one to spear, and the spearing knife was modified first into a two-tined fork, which was more effective in picking up food. They appeared as early as the seventh century in royal courts in the Middle East and were in use more widely among the wealthy in Tuscany in the eleventh century. The four-tined fork emerged later. The Fay paperclip, patented in 1867, was originally a competitor of pins and was used for holding material together but eventually settled into the paper niche. A variety of models were circulating late in the nineteenth century including the familiar "gem" paperclip whose design avoided scratching the user and seemed to also have more aesthetic appeal. It tended to out-compete other varieties except in very specialized niches and the even more practical "perfected" or square-ended gem evolved from it in the mid 1930s. These stories are told in more detail in Petroski (1992).

According to population genetics or the neo-Darwinian theory of evolution (the basic theory of evolution cast in a mathematical and genetical form), the first law of any evolutionary process is that the relative frequency of alternatives in an evolving population will remain constant unless some force like selection acts to change them. This is an inertial principle not unlike Newton's first law of motion. This "weight of history" is what social scientists and historians sometimes refer to as "path dependence". What takes place at any one time is, in part, dependent on what happened previously, sometimes far in the past, i.e. history is a cumulative process. Because this weight of history can sometimes frustrate adaptation, George Williams (1997) called it a "burden" which is demonstrable on many scales. Most of us are familiar with the lowered larynx which permits us to make a variety of sounds but also makes us more vulnerable to choking, or the retina with its blind spot where the optic nerve exits, or the way in which a relatively few body plans were laid out early in the history of multicellular evolution, and how, probably after some pruning, subsequent evolution has taken place largely within the framework of those body plans. Marcus (2008) has recently argued that the human mind itself is a "kluge" (rhymes with huge), an engineering term for a clumsy or inelegant solution to a problem for this reason. While this inertia, weight, or burden of history may sound like the "constraints" previously discussed, it is somewhat different in that the kind of internal "channelling" now under discussion is a result of the cumulative effect of the past history of evolution itself rather than universal laws or current correlations. The research programme of those who emphasize the role of history in evolution today is primarily that of systematics, particularly phylogenetic systematics or cladistics, which, without getting technical for the moment, classifies organisms uncovering the historical affinities among groups. This is discussed in Chapter 2.

The "adaptationist" research programme of evolutionary (including behavioural) ecology, on the other hand, emphasizes the power of selection rather than the weight of history. Evolutionary ecology seeks to complete Darwin's mission by providing not just a theory *that* selection is important in evolution, but a theory *of* selection in the sense of a theory of the conditions under which selection favours various kinds of alternatives. By comparison with systematics, evolutionary ecology is about similarity which is analogous or homoplastic (due to similar selection pressures) rather than homologous (due to a common history). Because it groups things occupying a similar ecological niche, it is sometimes said to study "guilds" (after the medieval "gilds" – associations of those in a similar craft) rather than "clades" (a common ancestor and all of its descendants). It seeks potential universal laws governing classes (e.g. gene pools, small things with fast life cycles) scattered across the tree of life over descriptions of historically unique lineages and their affinities (e.g. mammals) which are branches of the tree of life. It emphasizes the "modification" over the "descent" aspect of Darwin's description of evolution as "descent with modification". One of the great tasks for the future of evolutionary studies is a more complete integration of these two research programmes – controlling for historical causes of similarity in testing ecological principles and controlling for ecological causes of similarity in historical tree building. While the former is well underway, the latter has been hindered by the fact that general principles of evolutionary ecology are only now beginning to emerge. Some of these are discussed in Chapter 3.

1.5   DIFFERENCES IN EMPHASIS

While there is near universal agreement on the four factors – constraints, chance, history and necessity – not unsurprisingly, there is some, and in some cases much, disagreement over emphasis. Examples include the following. Molecular evolutionists disagree among themselves over the relative importance of chance in the form of sampling error relative to selection. Some molecular evolutionists are of the opinion that the majority of genetic innovations are "neutral" in their effect on fitness (viability and reproductive success). On a larger scale but somewhat similarly, the late Stephen Jay Gould once wrote a book on the Burgess shale fossils (1989) emphasizing chance combined with the historically cumulative nature of evolution (what he together called "contingency"), leading to his famous contention that if the tape of life were rerun, the results would be entirely different. To the contrary, Simon Conway-Morris (2003), covering some of the same ground, argued that a rerun would end up pretty much the same, including something very much like ourselves. Population genetics, in formally modelling evolution in genetical terms, does not normally explicitly mention history. Its first principle, however, is at least a minimally historical one – in the absence of forces causing change, existing relative frequencies of alleles (alternative forms of a gene) will remain inert through generations.

In Gould's last book (2002) chance was not much in evidence. He loosely referred to everything except selection as constraints which positively "channel" evolution as well as constrain it in a narrow negative sense. He divided constraints in this broad sense into two kinds. One was "structural" (the laws of physics and chemistry, as well as the physically/chemically necessary side-effects or byproducts of other changes – "spandrels") – what we have above called constraints (chapter 11). The other was historical (chapter 10). Here Gould rejected minimalist interpretations of the role of history such as the population geneticists' inclusion of only the relative frequency of alternatives one generation back. If what is observed at one point in time is partially a result of events in the distant past, then they also can have an impact far into the future. Technically, the argument is over the importance of parallelism. Groups with common histories may be "propelled" to evolve along similar trajectories (both because a common history makes it more likely that similar mutations and recombinations will arise and because a similar ecological environment subjects them to similar selection pressures). The concept of parallelism is a grey area between homology (similarity due to a common history) and analogy or convergence (similarity due to similar selection pressures) – in modern usage "homoplasy" includes both convergence and parallelism. The overall argument in both of his books and in both key chapters of his last book, however, is similar in one sense. While the main complaint of his earlier book was that conventionally too much emphasis is placed on selection and not enough on chance and history, more recently his complaint was again that too much emphasis is placed on selection, on the "functional", but now not enough on constraints and history.

Those who emphasize selection, on the other hand, are not silenced by such views. Constraints? Physics and chemistry to be sure but, as noted previously, correlations can be attributed to selection acting dependent on own condition as well as ecologically. Chance? To assume that rarity is piled on rarity is simply to multiply improbabilities into the realm of the absurd. History? There is no action at a distance in biology – events in the distant past affect those in the far future one step at a time through all of the intervening generations. Indeed, the primitive or ancestral (plesiomorphic) characteristics of a group have been no less subject to selection in the intervening period than have the more recently uniquely derived (apomorphic) ones – it is just that selection on the former has been negative, purifying or stabilizing (i.e. against change) while selection on the latter has been positive (for change). The reason for this can be that change in one character that would otherwise be ecologically adaptive would also have to be compatible with other characters and often it is not – creating a bias in favour of stability. Contrary to Gould, such views leave at best minor roles for constraints, chance and history, but a major one for selection.

Hence while there is variation in emphasis among evolutionists, it remains the case that almost everyone recognizes all four causes to some degree and, moreover, it is a fact that, in addition to population genetics which treats some of these formally, the two largest and most developed research programmes in evolutionary science – systematics and evolutionary ecology emphasize history and selection respectively. As a consequence of that, as well as of following Darwin's own "two great principles", we will follow that overall emphasis in this book. As mentioned in the Preface, as well as explaining basic elements of the process, this book emphasizes the breadth of applicability of Darwinian-style sociocultural evolutionary theory across the social sciences as well as the solutions that new developments not outlined in the foregoing sketch offer for solving dilemmas in contemporary social science.

1.6 A ROADMAP

Chapters 2 and 3, naturally enough given the above, deal with Darwin's "two great principles" – the "unity of types" (history) and the "conditions of existence" (selection) as applied to culture. Chapter 2 explains why the answer to the question, "Where did something come from?" is basically the same for cultural things as that Darwin proposed for living things, namely that they descend with modification in a tree-like pattern from other cultural things. This entails three claims about culture – that there is continuity, that there is change (the combination of the two meaning that things descend not only from the same, but also from different kinds of things) and that this change commonly takes place in a branching pattern. Continuity is fundamentally provided by social learning, learning by observation or instruction, which cannot be reduced to individual learning as Albert Bandura, the 2008 winner of the Grawemeyer Award, the most prestigious in psychology, first showed long ago (for overviews see Bandura 1971, 1977, 1986). Indeed, we now know that social learning by observation is present even in the animal kingdom, particularly among mammals and birds. The existence of this "second inheritance system" (Boyd and Richerson 1985) inevitably implies the existence of a second evolutionary system. In humans, cultural innovations like new variants of a religion develop in one or more individuals, spread by social learning and, importantly, spread differentially through a population under the influence of sociocultural selection. Individuals can and do make a difference. They sometimes produce strikingly novel and highly successful innovations. However, as noted in Section 1.3 above and as will be discussed in Chapter 2, there is no evidence in any area of human endeavour that, as a statistical body, innovations are biased in the direction that would be required for them to spread successfully. In fact, most fail. Sociocultural evolution, like the biological, is Darwinian rather than Lamarckian. After reviewing some of the evidence for continuity, change, and tree-like patterns, the chapter goes on to explain briefly how evolutionary theory answers historical questions about what is related to what and how their history should be reconstructed and proceeds to examples of how these methods have been applied in several social sciences.

Chapter 3 on "Necessity: why did it evolve?" begins with a discussion of three approaches to selection – the idiographic or historically unique; the nomothetic or lawful; and a third half-way house. At the one extreme is the historically specific approach, which maintains that so many factors are involved, in the final analysis, any particular instance of selection is unique. At the other extreme is the approach of population genetics, metabolic ecology and the cultural version of the former, which propose universal causal scientific laws. In between is the hybrid approach of evolutionary ecology. It seeks to complete Darwin's mission by providing a theory or theories of what selection favours under different ecological conditions wherever they may be located on evolutionary trees, and some principles that have emerged and sociocultural applications to the sociology of science are discussed. However, for now at least it is a hybrid in the sense that sometimes, when tested, its generalizations are found to hold in a universally conditioned way, but in other cases they are found to be confined to historically specific clades. The problem of "history versus science" is embedded deeply in the history and practice of all of the life and social sciences and according to some, in cosmology as well. However, there may be a greater opportunity to answer it in the life and social sciences than there is in the latter. Together, Chapters 2 and 3 make the case that "descent with modification" is a fact not a theory about the sociocultural world. The evolutionary approach of history *and* science enables us to reconcile the great early twentieth-century debate between Gabriel Tarde and Emile Durkheim over whether "imitation" or constraining social forces constitute the basic subject matter of the new discipline of sociology (for a summary see Clark 1969) and differences between cultural anthropologists and sociologists over whether culture or social relations/structure constitute the basic subject matter of the social sciences (for something of a truce see Kroeber and Parsons 1958).

Chapter 4 on "Competition, conflict and cooperation: why and how do they interact socially?" continues with the selection theme but moves on from the ecological to the social. Social relationships and interactions can act as a cause, a consequence, or both, of selection. Historically, economists have emphasized competition over strategic interaction. Sociologists and anthropologists on the other hand have emphasized social cooperation and conflict (traditionally in the form of functionalism and variants of Marxism, for example). Instead of either, this chapter adopts the biologist's three-cornered distinction between competition, conflict and cooperation. This helps make clear some general principles under which these alternatives should be expected to be observed as well as their likely consequences. Some of these principles are used to go into some depth on the nature of gender differences and relations. We will also see how the complexity of social relationships can be better appreciated by coming to understand the role of conflict in cooperative relationships and that of cooperation in antagonistic relationships. Finding the conditions favourable for the evolution of cooperation is among the most interdisciplinary endeavours underway today – engaging biologists, psychologists and social scientists with each other's theories and findings (e.g. Hammerstein 2003).

Darwin's reputation is that of the iconoclast who provided a materialist explanation of the history of life on earth and its present forms. Despite that, modern evolutionary theory with its emphasis on

information-related concepts (e.g. genes biologically and usually linguistically encoded information socioculturally) appears, in many
respects, more philosophically akin to idealism than materialism – a
dilemma that has plagued the history of the social sciences. Chapter 5
on "The ideal and the material: the role of memes in evolutionary social
science" will show that the common complaint that there are discrete
units of genetically encoded information which evolve – "genes", but
not of culturally encoded information – "memes", is not persuasive and
how in a number of cases, the "meme" concept has been useful scientifically. At the same time, much descriptive and evolutionary research
in both the life and the social sciences has also usefully been framed in
material "phenotypic" terms – suggesting that neither approach is
likely nor should be abandoned. The meme concept can be useful in
interdisciplinary discourse and with respect to cultural inheritance, is
most appropriate when social learning is by linguistic instruction, least
appropriate when it is by individual learning mechanisms, but the
situation is less clear when social learning is by observation. The
most reasonable conclusion to be drawn seems to be – memes if useful
but not necessarily memes.

Chapters 6, 7 and 8 on agency, subjectivity and complexity
respectively deal with three issues that sometimes tend to get confounded in the social science as all "micro versus macro" issues.
Chapter 6 is on "the problem of agency". Individual agency cannot,
with any scientific credibility, be understood simply as "free will".
Human psychology is not a black box and should not be viewed as
such by social scientists. The two most well-developed theories of
individual action apart from particular genetic and cultural influences
are considered. These are the most well-developed in the sense that one
possesses extensive empirical support and the other a complex, logical
deductive structure. Notably, the credibility of both are buttressed
by their real world applicability – the one constituting the foundations
of applied behavioural analysis, the other of microeconomics. The
relationship between the first "pragmatic" perspective (learning
theory) and the second "utilitarian" perspective (rational choice
theory) has been much discussed (e.g. Herrnstein 1990), but ultimately
we suggest that they are complementary. The latter is needed to
explain origins, and the former, which is shown to be a detailed evolutionary analogue, is needed to explain the mechanism and direction of
change in individual action.

For several decades now some theorists in a variety of social
science fields have been claiming that "everything is socially

constructed". This emphasis on subjectivity over objectivity has included fields such as science studies where the appearance of "relativism" has scandalized many "realist" scientists and philosophers. Hence Chapter 7 is on "subjectivity". Is everything, something or nothing socially constructed? One obvious point is that even if everything were "constructed", it would not necessarily always be "socially" so; something may be biologically or psychologically constructed. Moreover, both of these sciences have solved the fundamental conceptual issue involved. Recently, biological evolutionary theory has come to more fully appreciate than in the past that individuals and populations can evolve to *both* construct their niches and to be structured by them (Odling-Smee *et al.* 1996, 2003) suggesting a new definition of evolution by natural selection:

> Microevolution by natural selection is any change in the inductive control of development (whether morphological, physiological or behavioural) by ecology and/or in the construction of the latter by the former which alters the relative frequencies of (genetic or other) hereditary elements in a population beyond those expected of randomly chosen variants (Blute 2007, 2008a).

Similarly, psychologists have long have been reluctant to speak of "observation" and prefer instead to speak of "sensation and perception" (e.g. Wolfe 2006; Goldstein 2008). This is because they know that variation/change in "observations" can in some cases be attributable to variation/change among what is observed, in other cases to that among the observers, and in still other cases to some (any) proportion of each. The same balanced approach is available to the social sciences dealing with human groups. The implications of both perspectives in science studies, that on the one hand the institutions of science require defence but on the other hand that some weakening of credentialism and a greater opening up of participation in decision making in science would be a good thing both have merit.

   Chapter 8 deals with the evolution of complexity in the sense of both ecological complexity (more kinds) and individual complexity (more complex kinds). There we see how the physical, ecological or social environment can drive wedges into existing kinds, creating more kinds of things. While an old problem, the origin and evolution of individual complexity is one on which considerable and quite exciting progress has been made in the past decade or so (e.g. Maynard Smith and Szathmary 1995; Callebaut and Rasskin-Gutman 2005; Carroll 2005). Maynard Smith and Szaythmary conceived of this historically as the emergence of new kinds of aggregate individuals – of eukaryotic cells from the prokaryotic, of colonial and multicellular organisms from the unicellular, and of eusocial superorganisms from the multi-cellular, for example. The social sciences too face these kinds of questions about complex emergent kinds of things. For example, are formal organizations the kind of things which actually evolve? Perhaps instead they are only the kinds of things within and between which social roles or statuses evolve, or even only the cultural components of the latter in turn such as habits, routines or competencies evolve. Economists wonder if markets are so great, why are there firms? The evolution of new aggregate levels of replication – of truly recursive culture for example, is not only interesting in its own right, but because everything that evolves also develops, it has necessary impli-cations for understanding the mechanism of their development. In a discussion of "What is social structure?", Chapter 8 also suggests that the concept of social structure is more a polythetic or cluster concept rather than one for which necessary and sufficient conditions for its application are available. It tends to emphasize one side of many of the major issues in cultural and social theory discussed elsewhere in the book. When we ask, "Structure versus what?" that becomes clear. Social structure can be counter-posed to culture, change, history, func-tion, agency, construction and, of course, biology. Chapter 8 considers the micro–macro problem, in all of its evolutionary and developmental complexity but shorn of confounding with agency and subjectivity.

Dobzhansky once famously declared that "nothing in Biology makes sense except in the light of evolution" (1973). At least before this century is out, I believe it is likely that social scientists will similarly agree that nothing in social science makes sense except in the light of evolution either. However, those who already agree with the centrality of evolution to the social sciences do so with two quite different meanings in mind – the gene-based biological or, as has been emphasized here, the social learning-based sociocultural. Beyond either lies the incredibly complex tangle of how the two affect and interact with one another and with the psychological as well to pro-duce human behaviour, mind, culture and social organization. While some modest progress has been made – there remain more questions than answers. Chapter 9 on "Evolutionism and the future of the social sciences", which concludes this book, will comment on some of the issues involved.

2

History: where did something come from?

2.1 DARWIN'S TREE "I THINK"

"Darwin: The Evolution Revolution", a museum exhibit mounted by the American Museum of Natural History in New York, recently spent time in my home city, Toronto. It included a facsimile of Darwin's Notebook B, the first on transmutation of species – open at page 36 and also available online (van Wyhe 2002–9). The page includes Darwin's initial sketch of a diagram that was the only one to appear in the first edition of *On the Origin of Species*. It is a sketch of a tree-like structure, above and to the side of which he wrote: "I think". Below, the note reads as follows.

> Case must be that one generation then should be as many living as now. To do this and to have many species in same genus (as is) requires extinction. Thus between A and B immense gap of relation. C and B the finest gradation, B and D rather greater distinction. Thus genera would be formed. – bearing relation (continuing on the next page) to ancient types. – with several extinct forms. . .

While a bit cryptic, particularly with the diagram, the basic meaning is clear. Darwin thinks that there is an actual historical "great tree of life" as he called it. A and B are drawn as widely separated twigs, C and B as closely spaced ones, and B and D in between. He thinks that the significance of the "unity of types" is that species in the same genus are more closely related historically than are those in different genera, those in the same family are more closely related than those in different families, and so on up the "natural" taxonomic hierarchy of groups within groups. The basics of the natural system had been known since the many editions of *Systema Naturae* were published by the Swedish father of biological taxonomy in the eighteenth century, Carolus Linnaeus. Linnaeus viewed his system of classifying organisms in increasingly inclusive groups of species, genera, families, orders, classes and phyla as revealing the divine plan of God's creation. Darwin's quite different explanation and a later version of the diagram were laid out in more detail in Chapters 4 and 14 of the *Origin*. The diagram showed a stylized tree with the trunk at the base representing the common origin of life, the tips along the flattened top representing existing species, and tips that ended without making it that far representing extinct species. The explanation for why species naturally fall taxonomically into a hierarchy of groups within groups was supplied by the history of the branching patterns that gave rise to them.

Still, it was surprising how many aspects of evolution could not be encompassed by this tree metaphor and how many ambiguities there remained, many of which went unanalysed for a long time – but more on that in Section 2.6 below. Meanwhile the issue is whether the answer to the question "Where do cultural things come from?" is basically the same as that Darwin proposed for living things, namely that "They descend with modification in a tree-like pattern from other cultural things". That answer would entail three claims about culture: (i) that there is continuity, (ii) there is also change (the combination of the two meaning that things descend not only from the same, but also from different kinds of things), and (iii) that this change commonly takes place in a branching pattern. First let us consider a somewhat apocryphal story about the first claim – of cultural continuity.

2.2   THE US STANDARD RAILROAD GAUGE

The following story is one of those things that circulated widely on the Internet for a while in the late 1990s and for some years thereafter. I do not know whether it is true or not (more on that later). Meanwhile, if it is not, it should be – it is the kind of thing that often is true.

> "The U.S. Standard railroad gauge (distance between the rails) is 4 feet, 8.5 inches."
> "That's an exceedingly odd number. Why was that gauge used?"
> "Because that's the way they built them in England."
> "Why did the Brits build them like that?"
> "Because the first rail lines were built by the same people who built the pre-railroad tramways, and that's the gauge they used."
> "Why did 'they' use that gauge then?"
> "Because the people who built the tramways used the same jigs and tools that they used for building wagons, which used that wheel-spacing. Okay?"

"Why did the wagons use that odd wheel spacing?"

"Well, if they tried to use any other spacing the wagons would break on some of the old, long distance roads, because that's the spacing of the old wheel ruts."

"So who built these old rutted roads?"

"The first long distance roads in Europe were built by Imperial Rome for the benefit of the legions. The roads have been used ever since."

"And the ruts?"

"The initial ruts, which everyone else had to match for fear of destroying their wagons, were first made by Roman war chariots. Since the chariots were made for or by Imperial Rome they were all alike in the matter of wheel spacing. Thus, we have the answer to the original question. The United States standard railroad gauge of 4 feet, 8.5 inches derives from the original specification for an imperial Roman army war chariot. Specs and bureaucracies live forever. So the next time you are handed a specification and wonder what horse's arse came up with it, you may be exactly right. Because the Imperial Roman chariots were made to be just wide enough to accommodate the back ends of two war horses.

Now the twist to the story. When we see a Space Shuttle sitting on the launch pad, there are two big booster rockets attached to the sides of the main fuel tank. These are the solid rocket boosters, or SRBs. The SRBs are made by Thiokol at a factory in Utah. The engineers who designed the SRBs had to be shipped by train from the factory to the launch site. The railroad line to the factory runs through a tunnel in the mountains. The SRBs had to fit through the tunnel. The tunnel is slightly wider than a railroad track, and the railroad track is about as wide as two horses' behinds. So a major design feature of what is arguably the world's most advanced transportation system was determined by the width of a horse's arse."

Molotch (2003:106) repeats this story briefly. He attributes the first part of it to Hawken, Lovins and Lovins (1999:118) who unfortunately attribute it only to "one story". He attributes the second part of it to an Internet foreword supposedly coming originally from Howard Winsett at the NASA Dryden Flight Research Center. So we have come full circle back to the Internet meme. In any event, not surprisingly, the wide circulation of this text of unknown origin prompted a number of people to research the question. Also not surprisingly, their conclusions differed – ranging from basically true (Solomon 1998), to some of the specific claims are true and others false (Adams 2000), to basically false (Snopes.com 2007).

There is no doubt that this width (4 ft 8 ½ in) is known as the "standard" or "international" gauge in the industry, although narrower ones which are good for sharp curves and wider ones which are good for high speed are known elsewhere – so there clearly does exist variation. "Standard gauge" has been legally so in Britain since 1846 (influenced directly by George Stephenson's design for the Liverpool and Manchester Railway earlier and the subsequent recommendation of a Royal Commission) and came to dominate in the USA as well after the civil war. There is also little doubt that the pressure for standardization to facilitate trans-border transportation continues – globalization and all that. (Evolutionists call this "positive frequency-dependent selection" – it being advantageous to be like others, see Chapter 3.) That the width of the standard gauge can be traced back to Roman carriages is more difficult to verify. The author of the first source cited above refers to an "old edition" of the *Encyclopaedia Brittanica* in his possession and reports that Julius Caesar mandated the width after observing for himself the quality of a stone cartway at the Isthmus of Corinth in Greece. The author also reports that on a trip he personally confirmed the measurement at the excavated site. Interestingly, the third source above, which argues in the negative, does so on the same grounds that biologists adhering to the adaptationist programme discussed in the next chapter do – that stasis of this kind is not so much due to historical inertia as to functional considerations.

2.3   THE REDISCOVERY OF SOCIAL LEARNING

The railway gauge is a story of continuity. However interesting and entertaining, fortunately we do not have to rely solely on an Internet meme to document the reality of continuity in the realm of culture.

Darwin was lucky in one sense – that he did not have to deal with the first question of whether organisms come from others at all rather than arising de novo. It had for long been obvious that large plants and animals come from others, normally others of the same kind – horses give birth to horses, cows to cows. But only gradually from the experiments of Francesco Redi in 1668 through till Pasteur's experiments in 1864 was it eventually established that, even for small biological things, that also obtains – maggots do not come from mud, and the same is true even for organisms too small to be seen with the naked eye. Not surprisingly, it took time for the same thing to be clearly established for cultural kinds of things. Not two centuries to be sure, more like two decades once the specific question clearly arose – roughly from the early 1940s to the early 1960s when the micro-foundation of cultural continuity, the existence of social learning, was firmly established experimentally.

That story actually goes back further to the early twentieth century when the American psychologist and Columbia instructor Edward L. Thorndike formulated his "law of effect". Thorndike named a book he wrote about his experiments *Animal Intelligence* (1911, 2000) after a book of the same name by a young associate of Darwin – George John Romanes (1884). He did so because Romanes' book was full of anecdotes about the intelligence of animals which included, among other things, anecdotes about animals learning by imitation – a cat learning to beg by imitating a terrier (1884:414), a donkey learning to open a gate and then a barn door (1884: 333), possibly by observing people do it and so on. Thorndike just did not believe it and determined to find out exactly how animals did learn. Placing cats in latched cages with food outside, he observed what he thought was a period of random exploratory behaviour until the animal happened to hit on a way to open the latch. Thereafter, each time they were placed in the situation and successfully escaped, the length of time it took them to do so decreased. This decline in latency took place gradually. Thorndike concluded that the animals learned essentially by trial and error – the underlying principal of which he called the "law of effect". Reinforcement or reward "stamps in" neural connections between centres perceiving the situation and those responding to it. Thorndike's careful experimental work on learning led to a long and productive research tradition during which his cages evolved into automated Skinner boxes – a research tradition which continues to this day (and which is discussed more in Chapter 6). The tradition of the study of individual learning, however, also over-reached itself at times by claiming that individual learning mechanisms such as sensitization, habituation, Pavlov's associative and Thorndike's instrumental (or Skinner's operant) conditioning were the *only* mechanisms of learning.

In the 1940s, Miller and Dollard (1941, 1962) attempted to explain social learning by observation ("learning to do by seeing it done" as Thorndike described it) within this tradition. Their theory of "matched-dependent learning" proposed that learning by observation can be reduced to instrumental conditioning. The only difference is that, in the social learning case, the cue for doing x is someone else doing x – hence the description of the learning as dependent on reinforcement for "matching" one's behaviour to that of another. There is no doubt that this can take place – even in pigeons (e.g. Howard and White 2003)! But is it the only way, particularly in humans?

The next major advance was made by the social cognitivist Albert Bandura in the 1960s, stimulated in part by an interest in the effects of television violence on children (for an overview see Bandura 1971,

2007;1977). Among his most famous experiments were the "Bobo doll" experiments in which children exposed to a model exhibiting verbal and physical aggression toward the dolls subsequently exhibited both kinds of aggression at significantly greater frequency than those not so exposed. With these and a variety of other experiments Bandura showed that social learning (sometimes called imitation or true imitation) could not be reduced to operant conditioning. True social learning is confirmed particularly when the response is not in the previous repertoire of the learner, when the learner just observes not performs in the learning situation (and hence there is no possibility of reinforcement for performing at least), and when a long time lag exists between observation and performance (1977:36–37). Many years later, a study of citations, reputations, awards, etc., published in the *Review of General Psychology* showed Albert Bandura to be the fourth most eminent psychologist of all time after Skinner, Piaget and Freud (Haggbloom *et al.* 2002). In 2008 he was awarded the Grawemeyer award – the most prestigious and lucrative in psychology. (See also Section 5.6 on memes and social learning mechanisms, which provides a principled argument why social learning could not be reducible to operant conditioning.)

2.4   CULTURAL DESCENT IN HUMANS AND OTHER ANIMALS

Of course, social scientists have long been convinced of the existence of social learning – known to them under a variety of terms. While psychologists have traditionally spoken of social learning or imitation – sociologists have more often spoken of contagion (in crowds), diffusion (of innovations), collective behaviour (encompassing all kinds of rapid, small-scale social changes ranging from fads and fashions to social movements), socialization, and of course, writ large by anthropologists – culture. Not uncommonly, when social scientists have observed a behaviour increasing and then levelling off in an S-shaped (logistic) growth curve in a population (e.g. Rogers 2003), they have inferred, as did the Nobel prize-winning immunologist Macfarlane Burnet when studying the adaptive immune response for example, that "somewhere, something was multiplying" (1968:71).

The phenomena studied under these various labels have ranged from the trivial (coughing), to the simultaneously serious and amusing (collective hysterias and mass delusions) as well as the serious and unamusing (obesity, smoking) to the profound (suicides, happiness). We are all aware that laughing and yawning are contagious but are less likely to be aware that human yawns can be contagious to dogs (Joly-Mascheroni, Senju and Shepherd 2008)! With respect to coughing – in a classic study Pennebaker (1980) showed that in normal undergraduate lecture settings it can be contagious as well. Not surprisingly, the number of coughs per hour per person was highest in February and lowest in April and was higher during boring parts than in the more interesting parts of a film. Less obviously, the larger the group the more coughs per person – suggesting that coughing was contagious. However, he nailed the latter down by placing observers at nine points in the room and having them record coughs in their assigned areas with "coughograms" – records turned over every five minutes for a forty-five minute period. Sure enough, when someone in a sector of the room coughed, the more likely someone in that sector subsequently coughed. Coughing, in other words, spreads geographically though a room.

More serious, but also amusing, are the phenomena of collective hysterias and mass delusions. In hysterias, anxiety spreading in a small group in a confined space like a workplace takes the form of physical symptoms. By contrast, collective delusions are the "spontaneous, rapid spread of false or exaggerated beliefs within a population at large, temporarily affecting a particular region, culture or country" (Bartholomew and Goode 2000). The latter reviewed many of the most well known and documented cases of mass delusions from the past millennium. One of my favourites has always been the Seattle windshield pitting epidemic of 1954. Fuelled by fears over recent atomic tests, a few initial media reports that the windshields of cars had been "pitted" mushroomed over a couple of weeks into hundreds of calls to the police involving thousands of cars. At the height of the epidemic, the mayor of Seattle appealed to the President of the United States for emergency help. One early theory was that the pits were eggs laid by sandflies. Of course, as it was ultimately determined, people were just peering closely at the imperfections in, and tiny particles of, incompletely combusted coal on their windshields for the first time. A new one to me in the Bartholomew and Goode review was the epidemic of vanishing male genitalia in Nigeria in 1990. Many males became convinced that casual contact in crowds was being initiated to steal their penises. Raising an alarm and removing their clothes after an incident to check, they sometimes claimed that the shrunken (no doubt from fear) penis still there was not really theirs at all but must have been substituted! Serious and far less amusing is the evidence that obesity and smoking spread in large-scale social networks (Christakis and Fowler 2007, 2008).

Not at all amusing is the evidence compiled mainly by David Phillips and associates in the 1970s and 1980s for the social learning of suicidal behaviour. In some of the best sociology (1979, 1980, 1982, 1983; Bollen and Phillips 1981, 1982), general science (1977,1978) and medical journals (Phillips and Carstensen 1986; Phillips and Paight 1987) in the world, Phillips reported studies finding that suicides highly publicized in the media trigger suicides in the geographic area covered and that they increase in proportion to the amount of publicity. Not only that, but single car auto fatalities increase as well (the kind of cases police had long suspected might often be suicides). When murder suicides are reported, multiple fatality auto crashes increase. There is an age effect, and imitative suicide generally reaches a peak about three days after the publicity. Even fictional suicides can trigger such effects and even plane crashes can be triggered. These kinds of phenomena are so well established that responsible news outlets as well as operators of railway and subway systems have policies restricting publicity surrounding suicides.

To balance the horror of contagious suicide, it is also important to note that evidence exists that happiness is contagious as well. Fowler and Christakis (2008) employed methods of social network analysis longitudinally on twenty years of data from the Framingham Heart Study to show that happiness measured by standard means spreads through social networks of relatives, friends, neighbours and (less so) coworkers. They were able to show that happiness actually induces happiness in others rather than simply being a product of similar circumstances and that, remarkably, the effects can be observed up to three degrees of separation, e.g. on the friends of friends of friends.

I have emphasized the early social psychological proof of the existence of social learning and microsociological examples like coughing, collective hysterias, mass delusions and contagious suicide and happiness because many might not associate them with the concept of "culture" but that, of course, is what they are – subcultures – culture writ small. Social scientists talk about subcultures which are shared less extensively because everywhere we look, we find behaviour is governed by socially learned norms and values. Individuals are socialized into those appropriate to a particular role, status or social identity, even "deviant" ones. Most of us, however, are more familiar with culture writ large – the anthropologists' "way of life of a people" for which there are many detailed definitions. Almost all of these, however, emphasize that the culture of a people is "shared" or similar and is so, not because it is genetic, nor because it has been learned individually, but because it has been learned socially - i.e. members are similar because they share a common cultural ancestry. One of my favourite definitions is McGrew's (2003:433), that culture is "learned (rather than instinctive), social (rather than solitary), normative (rather than plastic) and collective (rather than idiosyncratic)". Another is Durham's (1991: Introduction) definition. To Durham, culture is first and foremost an "ideational phenomenon" or a "conceptual reality" - which is socially transmitted, symbolically encoded, systematically organized, and shared because of its social history. Notice, however, two differences. Durham's definition includes the cognitive nature of culture - culture is a matter of ideas and not simply of behaviour - while McGrew's does not. McGrew's definition includes the normative i.e. the prescriptive nature of culture - culture includes ideas about what one ought to do and think not just what is the case. (Counter-posing "normative" to "plastic" may be somewhat problematic depend-ing upon how "plastic" is interpreted. A norm is flexible in the sense that it is more general and therefore more flexible than a specified behaviour - we learn not just to say "please" and "thank you" but more generally to "be polite". It is not flexible in the sense that it is an "ought" rather than just an "is".) Both of these, the cognitive and the commonly, but not exclusively, normative in the sense of prescriptive nature of culture are important.

Biologists also now know that many animals other than humans learn socially by observation, in whatever sensory modality, and hence have simple cultures. (The only unambiguous case of learning socially by instruction in the sense of the transmission of symbolically encoded information, however, among animals is the famed waggle dance of honey bees by which worker bees communicate to others the direction and distance to food. The dance was discovered by Karl von Frisch in the 1940s, who shared a Nobel prize for his work in 1973, and has been confirmed by subsequent research - see, for example, Rilley *et al.* 2005). Some highlights in the history of the discovery of observational learn-ing in non-human animals include:

- experiments demonstrating that male chaffinches learn their song socially by hearing others in their first year of life (Thorpe 1958);
- the discovery of an old tradition (as judged by cavity wear) of nut cracking with hammer and platform stones at conventional sites among chimpanzees in west Africa (Sugiyama and Koman 1979);

- the publication by one of the great developmental biologists of his time, John Tyler Bonner of *The Evolution of Culture in Animals* (1980). While Bonner's analysis was of the steps involved in the biological evolution of the capacity for culture, he had no problem accepting that teaching and learning in non-human animals creates a new form of evolution – the cultural (e.g. p. 18);
- observations of the invention and spread of novel food gathering and handling behaviours among Japanese macaques on Koshima island including caramel eating, sweet potato washing and wheat sluicing (Nishida 1987);
- an experiment showing that nine-spine sticklebacks allowed to observe groups feeding at different feeders subsequently chose the one that previously delivered food at the higher rate (Coolen *et al.* 2003);
- the maintenance, by tradition, of different methods for solving the same (otherwise demonstrably unsolvable) problem, even with the same apparatus available, which were experimentally introduced by covertly training one individual in each of two chimpanzee colonies whose other members subsequently learned from observing them (Whiten, Horner and de Waal 2005);
- the subsequent experimental demonstration that such diffusion by observation can also take place between groups such as might occur in the wild when female chimps transfer between groups (Whiten *et al.* 2007).

While the emphasis in the study of observational learning in non-human animals has historically been on the role of the mimic rather than the model, there is increasing evidence that active "teaching" also takes place (for a review see Hoppitt *et al.* 2008).

In the last quarter of the twentieth century, despite such examples and many others, an attack on the idea that non-human animals learn socially by observation and hence possess culture was launched by some experimental psychologists, reminiscent of the earlier attack on the idea that humans do. This time the proposal was not based exclusively on the concept of matched-dependent individual learning but rather on breaking events down into pieces (e.g. Galef 1992 and references therein). For example, in a sequence of more than one behaviour such as locating food and then acquiring and consuming it, perhaps only the initial behaviour in the sequence is being learned

by observing another – "local enhancement". Perhaps in a single behaviour, only paying attention to, or perceiving, the stimulus is being
learned by observing another – "stimulus enhancement". In behavioural terms, "learning" is actually a complex of learning about four different things – an initial stimulus, a pattern of motor behaviour, a second
reinforcing stimulus (either before the motor pattern as in Pavlov's
associative conditioning or after it as in Thorndike's instrumental
conditioning), and an association between two or even all three of
these. In some cognitive interpretations, it may even be about learning
to bring about a particular result by a variety of alternative means that
are available – sometimes called semantic or programme level imitation, or emulation. Now while it may be of interest to experimental
psychologists which behaviour in a sequence, which component or
components of a single behaviour, or even whether a behaviour or a
programme, is being learned in a particular case – learning whatever of
these by observation is still learning by observation. Some experimental psychologists seem to have often been particularly concerned that
the inference was commonly being drawn unjustifiably, particularly by
naturalists who study behaviour in the field rather than in a laboratory,
that it was the pattern of motor behaviour that was being so learned
(sometimes called imitation or true imitation). Somewhat ironically, a
recent review of the literature concluded that there is in fact more
empirical evidence for "response facilitation" (learning a motor behaviour which may, however, have existed in the learner's existing repertoire) than for the other elements that have been proposed (for the
short and long versions see Laland 2008; Hoppitt and Laland 2008,
respectively).

Other issues which occasionally come up are irrelevant to the
basic one. Whether non-human animals are conscious, whether they
can behaviourally show that they "know what they know", or whether
the learning is cognitive in the sense that a programme and not just a
behaviour has been learned are different questions which also arise in
contexts other than social learning – in individual learning by trial and
error for example (on the latter issue there see Chapter 6). If a distinction is to be drawn among traditions, protocultures and culture – it
should be drawn in a theoretical and/or empirically meaningful way,
not in a way that distinguishes other animals and humans by definition. If all and only human social learning were normative, it would be
important. However, much human social learning is just about what is
and not about what ought to be, and social norms are commonly
enforced in other animal communities such as those pertaining to

dominance hierarchies for example. In general, what seems to be different in the social learning of humans and other animals is, first, the quantity and diversity of social learning by observation in the former relative to the latter and, secondly, the commoness of social learning by linguistic instruction in the former and its total or near total absence in the latter (although the current consensus that all other animals lack language may change as we come to understand animal communications systems better).

Overall, perhaps the best response to the criticisms of the exper- imental psychologists to the idea that non-human animals learn by observation was Frans de Waal's book *The Ape and the Sushi Master: Cultural Reflections by a Primatologist* (2001). Among other things, de Waal revisited the Koshima island site and talked to some of the researchers involved in the original discovery of food-related cultural traditions among Japanese macaques. As well as observing descendants of the original animals still washing and seasoning their sweet potatoes, he cleared up some distortions about what had taken place there which had entered into the literature (1998; 2001: chapter 5). It is notable that de Waal does much of his work in zoos. Those who work in zoos, and particularly those who work to reintroduce captive animals into the wild, are faced every day with the fact of social animals who often do not even know how to perform such basic functions as forage for food, mate, or care for their offspring unless they have been raised in natural communities in which these activities are observed from birth. Still, it is not surprising that the phenomenon of social learning by observation, whether in people or other animals, remains somewhat mysterious. After all, it is a special case of replication (or reproduction with heredity), and *biological* replication remains not fully understood. There will be more on this in Chapter 8 on the evolution of complexity.

2.5   WITH MODIFICATION IN HUMANS
AND OTHER ANIMALS

As well as continuity, evolution requires random variation and selec- tion resulting in change – whether linearly within a lineage (anagene- sis) or that resulting in a diverging tree-like pattern of dual or even multiple lineages (cladogenesis). As noted in the introduction, one of the most deceptively simple but difficult concepts for those coming new to evolutionary theory is to understand in what sense mutation or innovation more generally is "random" and hence that evolution, including cultural evolution, is Darwinian rather than Lamarckian.

First, "random" as used in evolutionary theory does not mean uncaused. There are many known causes of mutations including ionizing radiation, some chemicals and some viruses for example. Similarly, there are many known causes of behavioural changes in individuals – psychological learning processes of sensitization, habituation, classical and instrumental or operant conditioning for example. Secondly, it does not necessarily mean a unique or one-of-a-kind event. Biological mutation is often a recurrent affair – to the extent that the mutation rate from allele A1 to A2 can sometimes balance selection acting against A2, maintaining an otherwise maladaptive alternative in a population. Similarly, in well-controlled learning experiments – with pigeons in a Skinner box, for example – the application of the same learning procedures to different individuals will normally produce the same behavioural results. Socioculturally, a classic study in the sociology of science showed how commonly discoveries in science are "multiples" (Merton 1961). It should be noted, however, that in a broader context, the description of uniqueness often does apply, which is why the phylogenetic concepts and methods discussed in Section 2.6 below work. Darwin's and Wallace's concept of the major force causing evolutionary change was so similar that they agreed on the same term to describe it, "natural selection". However, there were great differences in their work more broadly – Darwin having marshalled many categories of evidence pertaining to geographical distributions, embryology and so on to support his theory – a fact that Wallace was always at pains to acknowledge. Thirdly, random does not mean that all possibilities are equally probable – "forward" e.g. from A1 to A2 and "back" e.g. A2 to A1 mutations commonly occur at different rates. More generally, mice mutate into slightly varying mice not into men. Fourthly, as noted in the introduction it does not mean that acquired characteristics are never inherited. If it does not mean uncaused, necessarily unique, equiprobable, or the complete absence of the inheritance of acquired characteristics – then what does it mean? As Campbell (1965) emphasized, the term "random" is ill-chosen. All random does mean in this context is "blind" or non-prescient. As also noted in the introduction, innovations are not necessarily oriented in the direction required for their successful spread. That is the case whether considering scientists publishing papers that will be cited, inventors obtaining patents that will be utilized, entrepreneurs founding new businesses that will succeed, or manufacturers introducing new products that will be successfully marketed (Blute 1979). On a statistical basis, most innovations fail and those that do succeed often do in a niche different from that which their inventor imagined. How, after all, could it be otherwise? It is not possible to know the future with any certainty until it is here.

Just as with continuity, change within a lineage in human culture has been demonstrated in phenomena ranging from the trivial to the profound. Stephen Jay Gould (1980) showed that in the short fifty years of its existence as a cultural icon, Mickey Mouse had evolved neotenously – a form of evolution in which descendants evolve to resemble the juveniles of ancestors. Specifically, by measurement no less, he showed changes in Mickey Mouse measured at three points towards a larger relative head size, larger eyes, and an enlarged cranium. He hypothesized the change reflected an unconscious discovery by the artists involved of the principle that such juvenile features evoke tender feelings of affection in humans akin to those evoked by our own babies. Not to be outdone, Hinde and Barden (1985) showed that teddy bears had similarly evolved in a neotenous direction presumably for the same reason. The original bear named after the American President Teddy Roosevelt had a low forehead and a long snout – it looked like a bear, whereas subseqently, there has been a trend towards a larger forehead and a shorter snout. Later, Morris, Reddy and Bunting (1995) found that the preference for infantile features in bears increased with age from age four to eight and was stronger in girls.

In the first half of the twentieth century at the same time as the study of individual learning processes dominated psychology, there also existed the field of social psychology which studied phenomena usually called "social influence". Once Bandura's work established the reality of social learning by observation – it became possible to go back and reinterpret many of the most well-known experiments in a common form as displaying transmission and selection bringing about population-level change. Among the experiments, none were more famous than those of Solomon Asch (1951) which showed that individuals asked to make various judgements about the length of lines will conform to the obviously erroneous judgements of others (about three-quarters of subjects did so at least once and about one-third of all judgements were erroneously conformist).

A common variant of the Asch experiment without victims and stooges and which has been repeated in innumerable psychology classrooms is the following. Students are asked to judge the length of a line displayed on a screen. One by one they give their answers aloud. As they do so the data are tabulated and displayed. In a large enough class it will take the form of a roughly normal distribution with the mean on or perhaps somewhat off the real length and the standard deviation being smaller or larger depending upon how difficult the task is made. Now comes the real experiment. With the distribution in front of them or at least having seen it, the students are asked to judge the same line again aloud and the data are tallied again. The result inevitably is that the mean stays about the same and the standard deviation decreases. The experiment is vaguely said to display social influence or conformity. However, what has actually taken place in detail can be understood in social learning or cultural transmission and selection terms. The two rounds of judgements are two cultural "generations". From the first to the second generation, the more extreme judgements in the first round have given rise to relatively fewer cultural offspring while those closer to the mean in the first round have given rise to relatively more cultural offspring. Students choosing the former have influenced few others (even themselves) while those choosing the latter have influenced more numerous others (including themselves). Cultural transmission plus selection has taken place. This form of selection in which extremes are selected against is called stabilizing or purifying by evolutionists but usually this is when selection stems from the ecological environment (i.e. here the line). In this case, however, the reason is as Asch and others have understood it – a tendency to conformity whatever its origin. This then is positive frequency-dependent selection, it being culturally advantageous to do what the majority are doing. Negative frequency-dependence is generally more common in evolution – it being advantageous to be unlike the majority because it reduces competition. However, it has been hypothesized that "conformist transmission", i.e. positive frequency-dependence, may have been important in human biological/sociocultural evolution with the result that similarity within groups and differences between them is increased – creating conditions favourable for group selection (Boyd and Richerson 1985; Richerson and Boyd 2005). (The evolution of cooperation and conflict is discussed in Chapter 4.)

Among the more profound cases of cultural descent with modification has been provided by George Basalla (1988) for the history of human technology. In case after case this distinguished historian of technology shows how human technological artifacts have descended with modification from previously existing artifacts. Nothing comes from nothing, everything comes from something. Do you think Eli Whitney invented the cotton gin, James Watt the steam engine, Joseph Henry the electric motor, Shockley Bardeen and Brattain the transistor, and Edison electric lights? In one sense of course they did – they innovated, making modifications and sometimes recombining previously existing things. But all had direct antecedents on which they built which themselves had antecedents, etc., back into the dim past (and indeed descendants as well). For example, Whitney's cotton gin which cleaned short staple cotton had antecedents in gins which cleaned long staple cotton, common even in the southern USA at the time. These, in turn, can be traced back through medieval Europe to the Indian gin or charka, which was in turn a variant of the older sugar cane press. Even imaginary machines have antecedents. George Orwell's imaginary book-writing machine in his novel 1984 descended from a similar imaginary device in Swift's novel *Gulliver's Travels*, which in turn was inspired by something real – children's alphabet blocks. Ultimately, Basalla goes so far as to hypothesize that all the material culture that surrounds us today has descended with modification from the first few stone tools used by human beings. And even before that there were antecedents – in nature. Barbed wire, invented in the American west to enclose livestock, was inspired by the thorn bushes originally used for the same purpose. Some general issues involved in understanding technological change as an evolutionary process are discussed in the essays in Ziman (2000).

O'Hara (1996) has pointed out how, not just descent with modification, but also historical tree construction has as old a history in the humanities (stemmatics of manuscripts) and social sciences (historical linguistics) (sometimes together known as "philology") as it has in biology. As is well known, Darwin appealed to the family tree of Indo-European languages as a metaphor to explain his view of the history of life. One of the great achievements of the social sciences has been the demonstration by historical linguists that the languages in each of the some two hundred human language families (containing some six thousand languages in total) have descended with modification from a common ancestral language (see any text in historical linguistics.) Some of these families in turn are historically related in superfamilies. Much more controversially, some consider it possible that all human languages on earth have descended with modification from a common ancestral language, sometimes called the mother tongue, proto-world or proto-sapiens (Ruhlen 1994a,b) and that the "click" languages of southern and eastern Africa may have been some of the earliest diverging branches of the universal tree of language (Tishkoff *et al.* 2007; but see also Güldemann and Stoneking 2008). For one version of the family tree of the world's languages see Figure 1, from Bichakjian (2002).

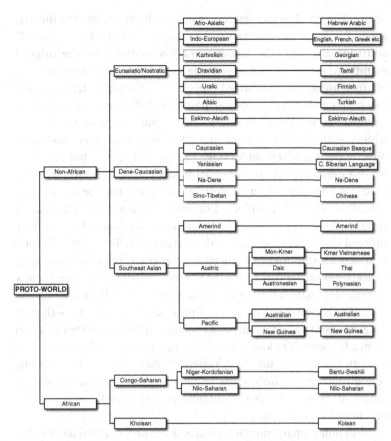

Figure 1. Family Tree of World's Languages. Reproduced with permission from Bernard H. Bichakjian, *Language in a Darwinian Perspective*. Frankfurt: Peter Lang Publishing (2002), p. 111.

Today, Darwinian-style cultural evolutionary theories of language change (such as those of Bichakjian as well as Croft (2000), Mufwene (2001) and Ritt (2004) – discussed again briefly in Chapter 8) are becoming the major alternative to the innate-grammar-based theories of Chomsky and his followers (Croft 2008:223; Evans and Levinson 2009).

Many familiar stories could be told of the branching history of human cultural phenomena, whether material or otherwise, and ranging from the micro to the macro. Essinger (2004) has told the story of how Babbage's original concept of his "analytical engine", essentially a general-purpose programmable computer never successfully built by him, however, was directly inspired by the punch card technology invented in France by Joseph-Marie Jacquard to speed up the weaving of silk. The hand loom therefore stands in direct ancestry to the first general purpose computers such as the Harvard Mark I and ENIAC. Subsequently, with the invention of microprocessors, the original mainframe computers have been able to proliferate into a whole series of increasingly miniaturized species – desktops, laptops, and touchscreen devices (including personal digital assistants, p.d.a.s). Astonishing symbioses have also taken place such as the iPhone, which, with its touch-screen interface, seems to have descended directly from the Newton Message Pad (an early p.d.a.) but also combines a camera-phone and portable media player with internet capabilities. At a more intermediate level, we could tell the story of the spread and diversification of organizations such as the parliamentary systems of government from seventeeth-century England. At the institutional level, anthropologists believe that the earliest human institution was kinship and that all others (religious, political, economic, educational, etc.) evolved from the former – presumably independently in many originally foraging societies. After all, people in hunting and gathering societies engage in all of these kinds of activities – they do religious, political, economic and educational things but they do them all in a framework of kinship statuses and roles rather than that of the modern kind of specialized institutions. Probably the earliest diverging role distinct from kinship was religious – shamanism in its various guises. At the most inclusive level there are the named human cultures of traditional anthropology.

In non-human animals too there is not only cultural descent but also modification. For example, when Whiten, Horner and de Waal (2005) seeded different problem-solving methods into two chimpanzee colonies and observed their subsequent transmission, that transmission was not without some change – they continued to diverge over time. The point was that the two cultural populations nevertheless remained distinct, signalling their different origins. Going back to the earlier work on bird songs, subsequent to Thorpe's experimental demonstration of social learning of the male song in chaffinches, change during the course of transmission in a population was directly observed in a variety of passerine "song" bird species in the late 1970s (e.g. Jenkins 1978). This led to the study of the cultural evolution of bird songs in which small variations introduced by accident in features like syllable, element or even whole song types and numbers are subsequently transmitted, resulting in the evolution of statistically different dialects in different local populations (e.g. Slater and Ince 1979; Slater, Ince and Colgan 1980). Some of the subsequent discussion involved analyses of whether these changes may primarily be due to cultural mutation and drift (Lynch *et al.* 1989). By 1980, the evidence for not just cultural continuity but for cultural evolution in non-human animals was sufficient for Mundinger to publish a major review emphasizing the evidence in birds and mammals. Today, the fact that cultural variation and evolution takes place in the songs of many birds is a textbook standard and is known in at least one case, and suspected in more, to play a role in their biological speciation (Catchpole and Slater 2008: chapter 9). And in some birds, it apparently goes beyond song to tool manufacturing. Hunt and Gray (2003) found that the three forms (wide, narrow and stepped) of a tool for capturing invertebrates shaped from Pandanus leaves by New Caledonian crows showed both cumulative evolution within each tradition, and geographic but not necessarily ecological overlap among them, suggesting that the three evolved culturally from a common ancestral form.

Roughly a decade after Mundinger's review, McGrew (1992) published a book laying out the case, mostly pro, for considering much chimpanzee interaction with the material world as culture pure and simple. In roughly another decade, a collective review by many of the world's leading primatologists documented the existence of 39 cultural variations among seven (Whiten *et al.* 1999) and later nine (Whiten *et al.* 2001) different geographic populations of chimpanzees not associated with ecological differences. In addition, evidence had accumulated that vocal and motor social learning and cultural traditions have evolved to be different in different Cetacean populations (whales and dolphins) of the same species with implications for conservation (Rendell and Whitehead 2001; Whitehead *et al.* 2004). We have now reached the point that a new sub-discipline "cultural primatology" (McGrew 2004) has emerged. As for any residual doubt by experimental psychologists about the field naturalists evidence for observational learning, i.e. cultural transmission in animals other than humans – the experiments described earlier by Whiten, Horner and de Waal (2005) and Whiten *et al.* (2007) have pretty much sealed the deal.

2.6 SYSTEMATICS REVOLUTIONIZED – PHYLOGENETICS AND THE NEW TREE OF LIFE

Section 2.1 above noted that there were things that could not be encompassed by Darwin's tree metaphor and ambiguities about others. Imagine we magically had a picture of the actual historical tree of life in front of us with dates arranged along a scale on the left ranging from the origin at the bottom to the present day at the top. Non-homologous similarities and differences, i.e. those not correlated with history but brought about by convergent and parallel changes or by reversals (together – homoplasies) as well as changes which might take place in a lineage rather than by their branching, cannot be represented by a tree diagram (unless one wished to employ horizontal distance regardless of the point of origin to represent this – in which case there might well be lines crossing all over the place). There are other ambiguities as well.

A single line branching cannot represent the complex and undoubtedly variable *process* of speciation. A species could range all the way from only one, very small local population, with very similar individuals, with each member of one sex having a virtually equal probability of mating with each of the other. At the other extreme, there could be many local populations, which differ substantially, with greatly reduced interbreeding between some or all of them, and with the degree of their similarity and reproductive isolation from each other waxing and waning through time. Indeed, even the degree of similarity and reproductive isolation need not necessarily be highly correlated with each other. Sibling species look very similar but are incapable of interbreeding with each other. In the opposite case, sometimes very differently appearing life cycle stages, morphs, genders, morphs within genders, etc., have been thought to be members of different species until investigated more thoroughly. Branchings and mergers are about the absence or presence of sexual interaction, while diversification and convergence are about the absence or presence of similarity. Moreover, in representing speciation, regardless of the details of the process which preceded it, Darwin's tree usually showed several branches going off from a single point. Was that consciously intended, and if so, why? Also, branches were never shown anastomitizing (merging). Why was that, given that Darwin viewed sterility as a matter of degree?

There were ambiguities as well about the relationship between classification and history particularly beyond the species level. Darwin distinguished between the classification and the related phylogeny or history that normally gave rise to it by using a different set of letters, but others did not always. Because some species speciate much more profusely than others, if classification is to remain true to history, some genera would include far fewer species than those crowded into other genera, for example. As well, major adaptive shifts undergone by some groups such as birds would be difficult to acknowledge if history were to count for everything and ecology for nothing. On the other hand, if classification were to more accurately describe the extant array of living things, then it might not be true to history. These questions exist at all hierarchical levels and, moreover, almost all agree that the number of hierarchical levels beyond the species that are recognized is rather arbitrary.

Three things eventually converged to bring about a veritable revolution in systematics and the effort currently underway to create a new tree of life. The first was the conceptual development of cladistics stemming from the work in the 1950s of the German entomologist Willi Hennig in his *Theorie der Phylogenetischen Systematik*. His work was not published in English, however, until the 1960s (Hennig 1966). Hennig had the goal of making classification strictly historical. Based on the insight that organisms are mosaics of characters – some very old, some less so, and some very recent, as well as the assumption that speciation takes place by bivariate branching, he pointed out that classification could be made strictly historical ("monophyletic") if it were based not on ancestral ("plesiomorphic") characters but on derived ("apomorphic") ones – specifically on *shared, derived* ones ("synapamorphies") – determined by the method of "out group" comparison. Hence, for example, humans and chimps share some characters that are *not* shared with gorillas and therefore they and their common ancestor form one clade (a common ancestor and all of its descendants). But then humans, chimps and gorillas share some characters that are *not* shared with orangutans and hence they and their common ancestor form a more inclusive clade and so on. Of course at what point a higher taxonomic group is recognized is arbitrary, but today the extinct species *H. neanderthalensis* and possibly *H. floresiensis* are recognized as belonging to the same higher level *Homo* genus as do we (*H. sapiens*). The great apes – chimps, gorillas and orangutans – are included in the same even more inclusive Hominidae family. If "rooted" (which end is which is determined), cladistic classifications can be used to construct phylogenies. (Accessible treatments of these concepts are available in sources such as Avise 2006; Lecointre and Le Guyader 2006, Introduction and chapter 1; Prothero 2007, chapter 5; and parts of the University of California Museum of Paleontology website – see particularly Caldwell and Lindberg 2009.)

A very roughly three-sided "systematics war" broke out in the 1970s and early 1980s over the ambiguities in traditional systematics – a story well told in David Hull's *Science as a Process* (1988) (but see also

Ebach, Morrone and Williams 2008, "cladistics of cladists"). Hull's sociological case study of the systematics community was presented in the context of his evolutionary theory of conceptual change in science. The three broad sides in the controversy were among those who, after Hennig, thought classification and nomenclature should be based on similarities and differences that are strictly historical (cladists), those who thought they should be based on overall observed similarities and differences (numerical taxonomists), and those who thought they should be based on a pragmatic mixture of both (who sometimes called themselves evolutionary taxonomists). The latter included probably the most prominent evolutionary biologist of his time – Ernst Mayr. While many of the general kinds of quantitative methods advocated by the second have become commonplace, and many taxonomists still function pragmatically, in the bigger picture there is little doubt that cladists have had the greatest conceptual influence. Most today agree that, while not always sticking to all of the principles of some of Hennig's more orthodox followers, cladism has nevertheless informed almost all modern systematics and made many changes in our understanding of the relationships among specific groups large and small.

The other two contributions to the taxonomic revolution were the coming of large amounts of molecular sequence data (DNA, RNA and protein) which required the use of quantitative methods, and of computers, algorithms and software programs capable of analysing the flood. Today the most common methods of constructing phylogenetic trees are "maximum parsimony" (the one most directly influenced by Hennig's ideas and which constructs all trees and chooses the one which requires the fewest changes in character states), "maximum likelihood" (a model-based method which, among other things, formally quantifies the relative probabilities of different trees) and methods based on Bayesian inference (prior or subjective probabilities). Three book-length overviews of the methods commonly employed in biology have been published in this decade (Semple and Steel 2003; Felsenstein 2004; Hall 2004) as well as a reader (Albert 2005). A comprehensive list and sources of software packages, many free, is maintained online by Felsenstein (2009).

All commonly used quantitative methods make classification and phylogeny construction more of a science rather than the esoteric arts they more traditionally resembled – ones in which the data and methods employed are laid out for others to see, criticize, revise, etc. In the Internet age it is suitable that knowledge about the evolution of life truly be made "public knowledge" in the form of the Tree of Life Web project (ToL 1996–2009). This collaborative project of biologists around the world aims eventually to include a page with pictures and information on every group of organisms, linked hierarchically. Other web-based projects are under development, such as the E. O. Wilson-inspired Encyclopedia of Life (EOL 2008–9), which describe biodiversity although not necessarily organized cladistically.

2.7 APPLICATIONS IN THE SOCIAL SCIENCES

As early as 1977 in the midst of the war over taxonomic philosophy and theory, and a good decade before Hull's (1988) sociocultural evolutionary study of the field was published, Platnick and Cameron were arguing for the generality of cladistic methods for studying historical subject matters including culture. As they summarized, "The concept that historical interrelationships can be demonstrated only by the presence of shared innovations is fundamental to the fields of textual and linguistic, as well as phylogenetic reconstruction. All three fields utilize analogous procedures. . ." (1977:380).

The extent to which these conceptual and methodological revolutions influenced the social sciences varied greatly by discipline and the kind of material being considered. The impact on the study of social organization in anthropology and sociology was slight. The study of formal organizations has been commonly carried out in modern times in business schools. McKelvey (1982) discussed the implications of the biological debates over systematics for the social sciences, particularly for the study of formal organizations. He developed a taxonomy of organizational forms in ancient Mesopotamia – work which went on to give rise to a large lineage of work on organizational evolution in general (e.g. Singh 1990; Baum and Singh 1994; Aldrich 1999, 2006; Durand 2006). Little of this subsequent work was specifically concerned with taxonomy, however, being more interested in questions such as the relationship between change within organizations and evolution among them (discussed in Chapter 8). Similarly, institutional economics, including the evolutionary, did not show a strong interest, being more concerned with questions such as the relationship between reason or agency and the evolution of economic institutions (discussed in Chapter 6). Generally, the influence has been strongest on those working with archaeological, linguistic and cultural anthropological materials. In 1971, Robert C. Dunnell published a book on *Systematics in Prehistory* beginning his journey towards becoming the founder of modern evolutionary archaeology. His own interest particularly developed in the direction of distinguishing selection (function) from drift (style) in the archaeological record and this interest has remained one in the field (e.g. Hurt and Rakita 2001; Brantingham 2007).

In the new century, the pace of social science applications in these fields accelerated. For example, O'Brien *et al.* (2001) applied cladistic methods to Paleoindian projectile points from the southeastern United States to develop a preliminary understanding of their evolutionary history. Gray and Atkinson (2003) applied maximum-likelihood methods and Bayesian inference techniques to a data set of Indo-European languages (87 languages, 2449 lexical items). Not surprisingly, the topology of the tree they derived more or less corresponds to the conventional one accepted by historical linguists who study the Indo-European language family. However, it was derived in a single quantitative study rather than by detailed analysis by hundreds of historical linguists over generations. In addition, the methods used were able to distinguish between two prominent theories of Indo-European origin. They found that rather than coming from nomadic Asian horsemen about 6000 years ago, the results support more recent theories that Indo-European arose from nascent farming communities in Anatolia (modern day Turkey) some 2000–4000 years earlier than that. Earlier, Gray and Jordan (2000) had applied parsimony methods to 77 Austronesian languages with 5185 lexical items and found it supported Diamond's "express train" model of Pacific colonization. It had proceeded in steps from Taiwan to central Polynesia in 2000 years, and Hawaii to the north, and New Zealand to the south were colonized from there. By the middle of this decade, sufficient research had accumulated in these fields for two edited volumes of archaeological applications to appear (O'Brien and Lyman 2000, 2003) and for Mace and Holden (2005) to review them. In 2005, Holden, Mace and Shennan edited a book of research papers from a conference written by some of the most prominent people in the field (see Holden and Mace 2005). The first half addressed the question of, "How tree-like is cultural evolution?" Of the nine papers in the section, one introduced the question and one explained and argued for the use of cladistic methods. Of the other seven studies, using a variety of methods, three on language families (Austronesian, Bantu and Indo-European) and one on material culture (Turkmen carpet designs) concluded that it is. Of two others on material culture, one (on a great variety of objects from the north coast of Papua New Guinea) found a mixture of vertical inheritance and horizontal diffusion and one (on indigenous Californian basket designs) found mainly diffusion between populations. (The second half was about the relationship between history and selection and will be discussed in the next chapter.)

More recently, Gray, Greenhill and Ross (2007) reviewed existing applications of phylogeny construction in all three cultural fields while Croft (2008:224–227) reviewed linguistic applications and Shennan (2008:80–82) reviewed archaeological applications. Gray, Greenhill and Ross's analysis of the growing literature showed that there are at least six fundamental kinds of questions about cultural topics, answers to which these methods can and have contributed. These are: (i) about "homelands", i.e. the geographical origin of something; (ii) about the order in which cultural lineages diverged and spread; (iii) about the timing of splits; (iv) about relative rates of cultural change; (v) about what something has evolved in adaptation to; and (vi) about ancestral states. They discussed the "perils" as well as "pleasures" of "Darwinizing cultures with phylogenies". However, they also point out that most of the issues which arise are not unlike those which also arise in the study of biological evolution, that there are methods available for dealing with them, and argue that ultimately, the proof of the utility of phylogenetic methods in the study of culture lies in the pudding of empirical research.

I am unaware of any applications of modern methods of constructing phylogenies to non-human animal cultural phenomena. Moreover, most of the applications to that of humans as noted lie in the general fields of archaeology, linguistics and to a lesser extent, cultural anthropology. Languages in particular, possesses plentiful data in the form of strings of discrete units. However, the possibilities should in no way be assumed to be necessarily confined to these. Various methods have been applied in literary studies (to manuscripts of *The Canterbury Tales* – Barbrook *et al.* 1998); in social anthropology (to descent systems, specifically patriliny versus matriliny among the Bantu and the relationship of the former with herding – Pagel 1994; Holden and Mace 2003, 2005); in the study of popular culture (to chain letters – Bennett, Li and Ma 2003); and in the study of historic rather than prehistoric material culture (to musical instruments – Tëmkin and Eldredge 2007) for example. Unusually, Tëmkin and Eldredge express more reservations than they do enthusiasm.

One of my favourite applications which Mace and Holden (2005) and Grays, Greenhill and Ross (2007) do not discuss in their reviews was in the field of scientific concepts – a cladistic analysis published by Stone (1996) of mathematical models of shell shapes in zoology.

I emphasize that this was a study, not of the shapes of shells, but of concepts in science – specifically mathematical models of shell shapes. Covering literature going back to 1838, Stone was able to distinguish nineteen species of models employed in this literature scored for the presence/absence of ten characters. The result of his cladistic analysis was two alternative incompletely resolved cladograms of the literature in the field which differed only in part. His phylogenetic interpretation was fascinating among other things for the light it threw not only on the conceptual history of the field (for example on the distinction between models which model form and those which model growth), but also on issues with respect to differences that may or may not exist in the processes of biological and cultural evolution. For example cross-lineage borrowing (homoplasy), anagenesis (change within an unbranched lineage), and historical constraints were present but uncommon and identifiable. Moreover the analysis was predictive. An Addendum added to the article quotes a referee who had been a participant in the research studied and which attests to the way in which the analysis uncovered an aspect of the history of the field which Stone, as a graduate student and outsider at the time, could have had no other way of knowing about.

While ideally phylogenies are constructed using quantitative methods, many of their basic features can be at least illustrated without them. Religion has become something of an evolutionary battleground in the past decade or so – evolutionism opposing intelligent design creationism, evolutionary theories of religion as a generally harmful parasitic memeplex opposing evolutionary theories of religion as a biologically and/or socioculturally group-selected adaptation and so on. Mindell (2006) largely avoided getting embroiled in these grand debates about religion. Instead, in his chapter five he simply illustrated how the logic of constructing phylogenies can be employed to understand the history of the Abrahamic religions. Circumcision was not unique to these religions, having been shared more broadly with other non-Abrahamic religions. Monotheism was a key innovation which remained stable for all the descendant religions although some of the accompanying innovations such as the dietary laws, Saturday sabbath, and Jerusalem as the main holy site were dropped within various branches of the Christian or Islamic traditions. The recognition of Jesus and Muhammad as prophets help define Christianity and Islam respectively. Mindell then provides more detailed phylogenies of the subsequent history of each of the Ancient Israelite Religion, of Christianity and of Islam. He shows, as did Stone

(1996) and Gray, Greenhill and Ross (2007), that many phenomena familiar to biologists are recognizable in the history of these religions. In this case these included the continued existence of early diverging groups (e.g. Samaritans, Jews of Ethiopia, Coptic Christians); convergence (e.g. in ethical ideals such as reciprocity); "adoptions", i.e. cross-lineage borrowings known as "syncretism" in the study of religions (e.g. the adoption by Jewish communities in south India of some Hindu customs such as a caste system and of not eating beef); and occasionally even promiscuous hybridization (e.g. Manicheism).

2.8 SUMMARY AND CONCLUSIONS

Darwin began a revolution in thought by interpreting the traditional Linnaean view that organisms fall naturally into a hierarchy of groups within groups, the "unity of types", as being the result of an historical process of descent with modification – one which can be represented by a tree. This revolution was deepened by Hennig's clarification of the significance of nested sets of shared, derived characters and by the development of formal methods of phylogeny construction such as parsimony, maximum likelihood and Bayesian inference.

The bulk of the social sciences, on the other hand, generally abandoned the idea that anything could be inferred about history by the comparative analysis of contemporary forms. Because their classifications were only one level deep and because, except for historical linguistics, they typically lacked the key concept of *uniquely* shared characters, i.e. of shared, derived characters, along with the appropriate formal methods of analysing them, their comparative method inevitably resulted in stage theories of historical progress. The conceptual roots of the latter in fact lie in the pre-Darwinian world of the medieval "great chain of being" (Lovejoy 1936) rather than responding to even the Darwinian (Blute 1979) let alone post-Darwinian innovations (Blute 1997). Inevitably, such theories were rightly rejected.

In taking a fresh look at this question today, few historically oriented social scientists would likely disagree with the fact that culture displays continuity with its microfoundations in social learning even in many other animals as well as change as discussed in this chapter. However, they might be tempted to respond, "So what if biologists, after Darwin, call these 'descent' and 'modification'?" They may even be aware that continuity with change, not uncommonly, takes place in a tree-like branching pattern as in the history of languages also discussed in the foregoing. However, a growing number of social scientists are additionally aware of just how much revolutionary concepts such as the significance of nested sets of shared, derived characters and formal methods of phylogeny construction such as parsimony, maximum likelihood and Bayesian inference can contribute to answering fundamental historical questions, including, but not confined to, what is related to what and where something came from. Despite potentially high rates of innovation and recombination, convergent evolution, reticulation and horizontal transfer of elements from larger wholes which have been discussed by many (e.g. Boyd *et al.* 1997), their work collectively has demonstrated that phylogenetic reconstructions of a great variety of cultural phenomena are possible. This in turn suggests that concepts and methods biological in origin could conceivably make a much greater contribution to historical methods in the social sciences (e.g. Hall and Bryant 2005), to concepts and theories in historical sociology (Adams, Clemens and Orloff 2005), to the new sociology of culture in general (e.g. Inglis and Hughson 2003), or in particular (e.g. Lieberson 2000 on names), and to cultural studies of material culture (e.g. Tilley *et al.* 2006) than they have to date. Now that one of the most outspoken critics of sociobiology, of gene-based biological evolutionary approaches to explaining human behaviour, Richard Lewontin, has turned his critical eye on meme or social learning-based sociocultural evolutionary approaches (Fracchia and Lewontin 1999, 2005; rejoinders by Runciman 2005a,b), perhaps that means that the approach has come of age!

I can do no better in concluding this discussion of "where does something come from" than to quote the late Charles Tilly (2006), one of the twentieth century's great historical sociologists. Less than two years before his death, in a didactic seminar on historical methods delivered to the American Sociological Association, he declared: "I predict a revival of epochal syntheses in sociolgy as biology's evolutionary *models* and findings become increasingly dominant in public discourse: why should sociologists let the world's Jared Diamonds monopolize the discussion?" (emphasis added).

3

Necessity: why did it evolve?

3.1 THREE APPROACHES TO SELECTION

In this chapter we turn from the first to the second of the two great principles of Darwin's theory of "descent with modification", from "the unity of types", i.e. history, to "the conditions of existence", i.e. natural selection. There are three broad approaches to understanding selection – the idiographic, the nomothetic, and a third evolutionary ecological approach which in some respects combines both.

Going back to history for a moment, an important thing to understand about taxonomy is that monophyletic "taxa" or groups of related organisms are historically specific entities, about which only idiographic (or particular) rather than nomothetic (or law-like) statements can be made (Ghiselin 1974b; Hull 1978a). At the species level for example, a human being is a human being not because of universally or even statistically shared properties with other human beings, but because of his or her physical connectedness to them. But according to cladistics, do not all members of a clade universally share one or more uniquely derived characteristics? Yes normally, but if these conflict, history trumps. For example, if habitually walking upright is one of the shared, derived characteristics which distinguishes human beings from the other living great apes, no taxonomist would claim that a person born with cerebral palsy and unable to walk is not therefore a human being. What matters is that a person's links can in principle be traced in a chain of physical connectedness – a chain of ancestor–descendant and/or sexual relationships to all other persons that have, do and ever will live. Because of this physical connectedness, spatio-temporal coordinates are implicit in the name of a species. Members of a monophyletic, or even a hybrid species for that matter, are not members of a class, such as those referred to in traditional scientific laws, but parts of a whole. A species is an "individual" as it is put – not in the sense of a functionally integrated organism, but in the sense that *H. Sapiens* is no less a proper name than is Charles Darwin. All have a specific origin in time, a specific spatial distribution, and a specific end in time. Just as is death for an organism, extinction of a species is forever. Some would have it that selection is similarly historically specific. The ability to survive and produce offspring is a product of the interaction between the character-istics of organisms and those of their environments. Both of these are so diverse that, in the final analysis, every case is unique. On that view, *all* of the study of evolution is an historical enterprise. Probably most of those who apply the selection concept to sociocultural entities think of selec-tion in that way. That is not, however, the perspective of those who take a nomothetic approach.

At the opposite extreme from the idiographic is the study of "classes" in the traditional philosophical sense – groups about which nomothetic (law-like) statements can potentially be made. Such state-ments are universal – "whenever and wherever", albeit they may be conditioned in innumerable ways as long as the conditions themselves are not spatio-temporally specific. All mechanical systems (approxi-mately) obey Newton's laws whenever and wherever they are found – not just in the USA between 1900 and 2000, say.

Evolutionists modestly call their laws "principles" like the Hardy–Weinberg equilibrium principle, the kin-selection principle and perhaps some of the principles of evolutionary ecology. The Hardy–Weinberg equilibrium principle for example states that the relative frequencies, i.e. the proportions of alternative alleles (different versions of a gene), in a population will not change through time unless some force like selection acts to change them. It is an inertial principle not unlike Newton's first law. In traditional law-like fashion, popula-tion genetics states the causal dependency of the relative frequencies of alternative alleles in a population beginning with their initial state, on a set of forces including mutation, migration, drift and selection (see any text in population genetics). The frequencies may change because mutation rates among them differ (e.g. A1 mutates to A2 at a higher frequency than A2 does to A1, for example), because one or another type migrates into or out of the population at different rates relative to others, because sampling error takes place in a finite population, or because of natural selection – differences in viability and/or reproduc-tive success. A slightly more complicated equation does the same for alternative diploid genotypes in a population whose members, like us, have two sets of genes, an equation which also includes a dependence on departures from random mating. Such equations apply to all populations with a Mendelian system of heredity whenever and wherever they may be found.

As far back as 1956, Gerard, Kluckhohn and Rapoport proposed that the basic principles of population genetics, including the role of mutation, migration, drift and selection are as applicable to purely cultural populations as they are to biological populations. They utilized language as their prime example and argued that a phoneme is formally identical to a classical gene. The approach was taken up most extensively by Cavalli-Sforza and Feldman in a series of articles and then a book (1981) on a quantitative approach to cultural transmission and evolution. They proved theorems involving the case of two alternative cultural states under vertical, oblique and horizontal cultural transmission as well as multiple state traits and those with a continuous distribution. Along the way, they applied the approach to many illustrative examples and data sets. The approach was extended by themselves as well as others – some of which will come up in Chapter 4 when considering the evolution of cooperation, in Chapter 8 when considering the relationship between cultural and social evolution, and in Chapter 9 when considering the concrete relationship between biological and sociocultural evolution. In this context, however, I wish to emphasize the traditional nomothetic nature of this research programme. For example, it maintains that the first law of cultural evolution is formally the same as the first law of biological evolution – the relative frequency of cultural alternatives in a population will not change unless some force like selection acts to change them.

Another ambitious but controversial nomothetic theory of the properties of organisms argues that their scaling properties can be derived from basic mathematical and physical principles. Although the facts remain in dispute, "metabolic ecology" or "the metabolic theory of scaling relations" claims that across species, metabolic rate tends to scale to the 3/4 power of body mass (log metabolic rate is a linear function of log body mass with a slope of 3/4) – a phenomenon called Kleiber's law. Logarithmic or power functions relate rates of change in variables to each other in a linear fashion and the 3/4 slope is somewhat different from the 2/3 conventionally thought to obtain and which would be expected on simple geometric considerations. (Geometrically, surface area increases with the square of length whereas mass increases with its cube.) The theory proposes that the 3/4 ratio can be derived with fractal geometry and fluid dynamics as a solution to the problem of minimizing the energy required to deliver resources in a branching network (e.g. blood vessels, xylem) to the maximum number of similarly sized terminal units that can be packed into a three-dimension space (West, Brown and Enquist 1997; for popular summaries see Whitfield 2004, 2006 as well as Math Bench 2009). The approach has been extended to myriad scaling problems based on quarters rather than thirds to the point where it is sometimes, only half jokingly, called for biology "a theory of everything". To my knowledge, metabolic ecology has not penetrated human cultural, political or economic ecology. For example, one of the most synthetic treatments of which I am aware that deals with energy acquisition and utilization in the history of both life and human societies (Smil 2008) is not based on West, Brown and Enquist's (1997) packing and transport theory.

Despite their great generality, both population genetics and metabolic ecology appear to be impoverished theories to "splitters" as opposed to "lumpers". In population genetics the ecological environment is present in the sense that it is what is responsible for the selection term, but present in a grossly inclusive way. In the case of the metabolic theory, the ecological environment is absent in the sense that all of life is said to be solving the same transport and packing problem. As noted in the introductory chapter, some seek not just a theory *that* natural selection is important, but a theory *of* natural selection in the sense of a theory of the conditions under which selection favours various kinds of alternatives. This is the third "adaptationist" programme of evolutionary (including behavioural) ecology such as that of life histories. By comparison with systematics, evolutionary ecology is about similarities which are analogous or homoplastic (due to similar selection pressures) rather than homologous (due to a common history). By contrast with historical clades, it is sometimes said to study ecological guilds (after the medieval "gilds" or associations of craftsmen in a trade). It emphasizes the "modification" over the "descent" aspect of Darwin's description of evolution as "descent with modification". It is akin to the nomothetic programmes of population genetics and metabolic ecology, not in the sense that it formulates statements about all populations anywhere and everywhere, but in the sense that it formulates ones in a more conditional but still nomothetic style. The entities studied like large and small clutch sizes for example are not historically specific branches or twigs of the tree of life, but rather are universally specified classes scattered across the tree of life. However, as we shall see, despite its ambitions, so far at least, it is a kind of half-way house between the idiographic and the nomothetic because it is often forced to condition its generalizations

by taxonomic group, and these "groups" of course are historically specific entities, those branches of the tree of life discussed earlier. But
before going on to discuss this hybrid programme, the dilemma involved
is well illustrated by Basalla on extraterrestrials.

3.2   BASALLA ON EXTRATERRESTRIALS

In the previous chapter we discussed briefly the historian George
Basalla's book on *The Evolution of Technology* (1988). There he analyses
the history of technology as a product of an evolutionary process
including continuity, novelty and selection. We specifically mentioned
the many cases where he demonstrates the reality of descent with
modification in technology. His more recent *Civilized Life in the
Universe: Scientists on Intelligent Extraterrestrials* (2006) is a gem of a book
as well. Not surprisingly for an historian, Basalla comes down on the
side of history over necessity, the idiographic over the nomothetic.
However, the book well illustrates why one might be pushed in one
direction or the other. On the question of the existence of civilizations
elsewhere in the universe there are believers and sceptics – agnostics
are few and far between. Basalla is a sceptic, but never explicitly. Like a
lawyer marshalling his case minus opening and closing arguments, in a
flat dry voice, he simply tells the story of speculation by philosophers
and scientists from the ancient Greeks right up to SETI and beyond –
often letting his subjects condemn themselves in their own words. Like
Sergeant Joe Friday of Dragnet, Basalla himself gives us, "Just the facts,
Ma'am, just the facts".

Speculation about the existence and nature of extraterrestrial
civilizations has been based on a trio of three key ideas – that the
universe is very large, even infinite; that there are other inhabited
worlds; and that other civilizations are superior to ours (Basalla calls
the latter "the principle of mediocrity"). In all periods, the speculation
has been fuelled by religious impulses and anthropomorphism, however
buried and, more often than not, not so deeply buried at all. As one reads
of case after case of otherwise intelligent philosophers and scientists
getting lost in a world of their own fantasies replete with lavish details –
of Kepler on the Subvolvans, the Privolvans and the savages inhabiting
the Moon; of Herschel on Martian cities and villages; of Schiaparelli and
Lowell's maps of Martian canals; of Crick and Orgel's directed pansper-
mia (the seeding of life on earth directed by intelligent aliens); of Dyson's
spheres built by aliens to surround a star and its planets; of Kardashev's
detailed description of three types of extraterrestrial civilizations; of

Sagan's interpretation of the "fish" origin myth of Ancient Sumerian civilization as possible evidence of previous contact – the reader cannot help but react exactly in the way Basalla intended, "What *were* they thinking!" Nowhere is the postmodern constructionist case for the subjectivity of a field of science made any better (and without even mentioning that dreaded pair of words).

While not as explicitly evolutionary as was *The Evolution of Technology*, there is a deep connection between the two. The evolutionary process (whether biological or sociocultural) and our understanding of it is Janus-faced. As noted in the previous section, on the one face we have history – systematics for example classifies organisms (or cultural kinds of things) and reconstructs their histories (phylogenies). The things studied – monophyletic taxa or clades, (i.e. a common ancestor and all its descendants) are historically specific kinds of things. On the other face we have science as more conventionally understood – proposed universal laws (typically called "principles" in biology such as the Hardy–Weinberg equilibrium principle, the theory of metabolic ecology or possibly some of the evolutionary–ecological principles of life history evolution), principles which act on populations, depending upon conditions of course, scattered across the tree of life. It is this distinction between an emphasis on the idiographic and the nomothetic that structures the attitude of believers and sceptics with respect to the plausibility of extraterrestrial civilizations.

Believers emphasize the universality of mathematics, of the laws of physics and chemistry and the advantages of sophisticated technologies and complex social organization. Consequently, they infer that extraterrestrial civilizations are probable. Sceptics emphasize the historically specific nature of the patristic products of biological and cultural evolution. Consequently they cannot imagine why one would think it likely that something even remotely like ourselves would exist elsewhere. Basalla's otherwise sophisticated evolutionary interpretation of the evolution of technology in his previous book implicitly emphasized the historical face of evolution. He interpreted selection in the history of technology as stemming from a whole plethora of diverse economic, military, cultural and social factors for example. So it is not surprising that he has landed among the sceptics on extraterrestrials. It should also not be surprising, however, if others, particularly the physicists, astronomers and cosmologists who have tended to dominate the subject, but even other evolutionists gazing on the other face of Janus, have come to a different conclusion.

3.3   SOME PRINCIPLES OF EVOLUTIONARY
ECOLOGY INCLUDING IN SCIENCE

The evolutionary ecology of life history characteristics has been an active field of research (for book-length treatments see Stearns 1992; Charnov 1993; Roff 2002; Bonner 2006) and is also commonly discussed in monographs and texts on behavioural ecology (e.g. Danchin, Giraldeau and Cézilly 2008). It borrows the concepts of scarcity, constraints, optimality and implicitly spending and investing from economics. It assumes (i) trade-offs, i.e. resources devoted one way cannot be devoted in another and vice-versa; (ii) constraints, i.e. the ecological environment enables and constrains what can evolve; and (iii) that natural selection optimizes fitness, however defined. It typically (although there are exceptions) formulates theories in phenotypic rather than genetic terms which is justified under rather restrictive conditions – particularly that the many genes involved in influencing such complex phenotypic characters as life history characteristics are not linked and that their effects are each small and combine additively.

What lurks behind many of its more specific theories is the assumption that the basic physical features of the ecological environment (energy content, time and space – both literal and metaphorical, i.e. niche time and space) are the properties that evolving organisms are most likely to evolve adaptations to. In that sense it is built on physics as well as economics. Because one of the most well-developed applications of evolutionary theory to a sociocultural phenomenon was Hull's (1988) application to conceptual change in science, Blute (2003) extended this to consider life history evolution there and what follows on this topic is modified from that.

We assume, as does the sociology of science in general, including Hull's, that scientists compete for status or "credit". They compete to acquire knowledge through research and to produce cultural heirs or descendants both by means of graduate students (somewhat similar to asexually produced offspring) and through publishing (somewhat similar to sexual gametes). In the latter case, the results of their work get combined with that of others in the minds of readers, incorporated into their research, and eventually into their graduate students and publications, with credit acknowledged by citations. We begin with density-dependence but because ecological factors controlling sizes and numbers may differ, we consider consumption (research) and production (teaching and publishing) separately.

3.3.1 Density

Density-dependent selection was first discussed in the context of r and K selection theory (MacArthur 1962; MacArthur and Wilson 1967; Pianka 1970) – discussed in an appendix to this chapter. Assume here that the amount of environmental resources available vary but that spatio-temporal boundaries including the metaphorical, i.e. niche boundaries are fixed. At or with a history of low population densities (measured in size, i.e. by the amount of knowledge possessed by the average researcher in the field relative to that available to be discovered) conditions are benign. Selection should favour productivity or "primary" research. This involves essentially spending on feeding, acquiring as much knowledge as possible – collecting data, developing new methods, adding axioms or postulates to existing theories or new theories themselves. At or with a history of high densities, however (i.e. once the average competitor possesses a lot of knowledge relative to that available to be discovered in a field), conditions are difficult. Selection should favour efficiency or "secondary" research. This involves essentially investing in digestion, deriving more "break down products" from each unit of knowledge acquired – reanalysing data, refining methods, proving theorems using existing axioms, postulates or theories etc.

Generally, in biology smaller organisms should excel at eating and the large at digesting. The surface area of a sphere $= 4\pi r^2$ while the volume $= 4/3\pi r^3$. Hence the ratio of $A/V = 3/r$ so that as a sphere becomes smaller (r approaches 0) the surface to volume ratio approaches ∞, and as it becomes larger (r approaches ∞) the surface to volume ratio approaches 0. Hence small organisms have a proportionately greater surface area for their volume (and at constant cytoplasmic densities, for their mass) than do large ones – surface area which can be utilized for feeding, sensing food, etc. Conversely, large organisms have a proportionately greater volume (and at constant cytoplasmic densities, greater mass) for their surface area than do small ones – volume and mass that can be employed for complex internal digestive processes.

Analogous principles apply demographically. At or with a history of low densities (measured now in numbers of researchers competing in a field relative to positions available or the number of papers published relative to readers of them) conditions are benign and a whole field is available to be populated by researchers/publications. Under such conditions, selection should favour quantity or "primary" teaching and publishing – spending on producing as many graduate students and publications as possible. At or with a history of high densities, on the other hand, competitors are many and there is not room for many more. Under such harsh conditions selection should favour quality or "secondary" teaching and publishing, investing in deriving more grand-offspring, i.e. "grand-students" and "grand-papers" from each of the fewer students and papers produced. It is surprising then to find that, in science, longer papers get more citations, as has been reported (Stanek 2008, cited in Ball 2008). One might expect them instead to get fewer citations but then the few papers that cite them to garner more citations in turn, i.e. "grand-offspring". One possible explanation of the unexpected finding is that it has been shown through the analysis of errors that many citations are not taken directly but are instead copied from other papers (e.g. Simkin and Roychowdhury 2002, cited in Ball 2002). Hence the higher citation rates of longer papers would be explicable if some of those citations were copied from others and hence the papers were actually grand-offspring rather than offspring.

Note that within either consumption (research) or production (teaching/publishing), neither of the strategies expected at low versus high densities is intrinsically better than the other – rather, both should be similarly successful in their appropriate circumstances. The FASEB (Federation of American Societies for Experimental Biology) released a report in mid 2007 summarized by Check (2007) showing that selection is not optimizing in graduate education in the biomedical sciences in the USA. In roughly the preceding decade, the percentage of PhDs with tenure or tenure-track jobs dropped from nearly 45% to below 30%. Similarly, the success rates in obtaining NIH postdoctoral fellowships declined from a high of around 45% to a low of just over 25%. Why then have scientists not shifted to producing fewer, higher quality graduates? Is it just a lag? Actually something else is probably going on, a clue to which is that most NIH-supported postdocs are now being funded from research rather than training grants. This reflects the emergence of a primitive colonial or even multicellular form of social organization in which the hierarchy of postdocs, doctoral students, undergraduate research assistants and so on working in academic labs are there to feed the principal investigator and the few successful postdocs and PhDs which do emerge from them – they are "disposable soma" or "hands in the lab" as the saying puts it.

3.3.2 Scale

Density-dependence as used above assumes that the amount of resources available varies but spatio-temporal boundaries are fixed. If we switch controls and assume instead that spatio-temporal boundaries vary while the amount of resources available is fixed then we can distinguish small-scale environments in which the energy content is more concentrated from large-scale ones in which it is smeared out in time/space or niche. Small-scale environments should favour r/d (for rate/density) strategies while large-scale ones should favour t/s (for time/space) strategies, i.e. selection should be scale-dependent. In science, effort may be devoted to spending on doing research rapidly but for a shorter period of time in a field, yielding a higher rate (mass-per-unit-time so to speak) of knowledge. Alternatively, effort may be devoted to investing in doing research at a slower rate but for longer (yielding knowledge over a longer time period). Trade-offs on a fast–slow continuum(s) are among the most well-known and well-documented in the biology of life histories (Read and Harvey 1989; Promislow and Harvey 1990; Bielby *et al.* 2007). "Live fast, die young" is the rule in Promislow and Harvey's (1990) memorable phrase. The equivalent distinction can be applied spatially including to body size, home range or metaphorically to niche breadth. The latter is familiar sociologically. In science, effort may be devoted to spending on doing research more intensively but in a smaller area of a field (yielding a higher mass-per-unit-area of knowledge so to speak), i.e. as a "specialist", or alternatively, less intensively but investing in spanning a larger area, i.e. as a "generalist". Similarly with teaching and publishing – effort may be concentrated narrowly on a single topic or spread more thinly but broadly across a number of topics. Organisms as well as scientists in their research and teaching/publishing should evolve to be adapted to the literal or metaphorical spatio-temporal scale of environmental opportunities, whether small or large.

Moreover there seems to be a correlation between spatial and temporal strategies so that smaller organisms (e.g. measured in length or with smaller home ranges or who occupy more specialized niches) tend to have fast life cycles, while the larger (e.g. longer, or with larger home ranges or broader niches) tend to have longer, slower life cycles. The reason for such correlations is not obvious – to attribute them to ecological correlations simply moves the need for an explanation one step back. I have suggested that they may exist because both r and d spending strategies are expensive in "labour power" (in biology ATP or mitochondria) while both t and s investing strategies are expensive in "capital" (in biology enzymes which are tools, inventories of proteins and other substances which are building blocks, ribosomes which are protein factories, and mRNAs which are managerial knowledge and expertise). Hence individuals may have a comparative advantage for one or the other strategy combination. Selection is known to be not only ecologically, but also own condition-dependent. In science, individuals may simply be differentially predisposed by temperament, however acquired, to be fast specialists or longer, slower generalists.

3.3.3 Patchiness

If we now assume that *both* the amount of resources and the boundaries are fixed, it can still be the case that those resources are more continuously or more discontinuously, i.e. patchily distributed in time, space or niche within those boundaries (evolution in changing environments was first discussed by Levins 1968). In such a patchy environment, if *local* densities are low, selection should favour spending on acquiring them (here including digestion). If they are high, then selection should favour investing in "moving" – whether in time (diapause, i.e. dormancy or hibernation), in space (motility or migration) or in niche (innovating even blindly, to change niches – what I have called phenotypic mutability but some might call investing in evolvability) – collectively the "3m"s. Note that in a finite world, investing in motility implies eventually recolonizing the same patches – a strategy called opportunistic among animals and shifting or swidden cultivation among human agriculturalists. Parallel distinctions apply to production – spending on producing more offspring versus investing in producing ones which are resistant, motile or which are phenotypically innovative. In either the consumption or production case, which of the three investing strategies is favoured should be determined by whether the environment is patchy in time, space or niche respectively – and more generally by their "net present value". Business people understand these alternatives. While their objective is always to grow their business, sometimes this requires investing in waiting for a better time; moving elsewhere such as overseas; or innovating, changing the market competed in, in the hope of landing in the right one. What scientist has not put aside a research project on hold for a better time, moved their job to a new geographical location, or changed fields of research? Similarly, what scientist has not temporarily extended support for a graduate student or postdoc until a suitable position for him or her is available, trained one who moved overseas, or trained some for a somewhat different field than others?

These spending versus investing strategies favoured under different local density conditions are different from the scale dependent r/d

versus t/s ones. This is because in this case there is a gap between resource sub-patches or items. In a seasonal environment, a perennial plant which grows and then at the end of the growing season whose foliage dies back in dormancy, or an animal which feeds and then at the end of the feeding season hibernates, is not switching to consuming over a longer time period; it is passing through a temporal gap without feeding until the environment is renewed again. Similarly, an animal which feeds, and then when resources become scarce migrates, is not switching to feeding from a larger home territory; it is passing through territory not feeding until it reaches a new place where resources are available again. The same applies to the distinction between spending on producing offspring versus investing in offspring which are dormant, motile or are innovative.

3.3.4 Stochastic environments

Thus far we have assumed evolution is taking place in an unchanging, unvarying environment (Roff 2002, chapter 4). Population densities were either low or high, environmental scales either small or large, or in patchy environments, local population densities were either low or high. However, conditions may be stochastic, i.e. predictable only in probabalistic terms (Roff 2002, chapter 5) which generally favours "bet-hedging". Like choosing an index fund rather than stock picking or market timing, under uncertainty conservatively doing some of both alternatives at issue in whichever case yields a higher geometric mean fitness (the nth root of the product of their fitnesses rather than their sum divided by n), i.e. a reduced risk of extinction. Particularly interesting is the case in which, despite uncertainty, reliable cues are available (sometimes called predictable conditions, Roff 2002, chapter 6). Under these conditions – investing in flexibility ("adaptive phenotypic plasticity") is favoured – investing in the capacity to vary and/or change developmentally, whether morphologically, physiologically or behaviourally, in a way that is responsive to, and adaptive in, the circumstances present. As a consequence, the same genotype can give rise to varying and changing phenotypes – whether within an individual/generation (e.g. the prey and patch models of classical foraging theory) or even between them (e.g. different morphs or generations of dual or even multigenerational life cycles, the latter possible presumably because of epigenetic inheritance). If beneficial, scientists too often prove themselves adept at being flexible – varying and changing among being productive or being efficient, between working as fast specialists or as

Table 1. *Four fundamental ecological dimensions and the strategies favoured by selection in science*

Ecological dimension[a]	Condition 1	Strategy favoured	Condition 2	Strategy favoured
Population density	(i) Low in size (ii) Low in numbers	(i) Primary research (ii) Primary teaching/ publishing	(i) High in size (ii) High in numbers	(i) Secondary research (ii) Secondary teaching/ publishing
Environmental scale	Small	Fast, specialized research/ teaching/ publishing	Large	Longer, slower more general research/ teaching/ publishing
Environmental patchiness	Low local density in size/ numbers	Spend on research/ teaching/ publishing	High local density in size/ numbers	Invest in waiting, moving or changing fields in research/ teaching/ publishing
Stochastic environments	Uncertain	Spend on bet hedging	Uncertain but with reliable cues	Invest in flexibility

[a] Population density: assumes amount of resources available varies, boundaries are fixed. Environmental scale: assumes amount of resources available is fixed, boundaries vary. Environmental patchiness: even with both fixed, *local* population densities may be low or high. Stochastic environments: the conditions encountered by an individual can be stochastic rather than deterministic but reliable cues may also be available. Also to be considered are antagonists, and the relationship between consumption (research) and production (teaching/publishing). See text for full explanation.

longer slower generalists, between staying put or moving on and so on in ways that further their careers. The four ecological dimensions and sets of alternative strategies favoured under different values of them which have been discussed so far in this section are summarized in Table 1.

3.3.5 Antagonists

Thus far we have considered the selective importance of ecological resources rather than antagonists like parasites and predators, but these are also important in evolution. Similar principles can be applied but generally in reverse with respect to antagonists. Hence for example as just discussed, in a patchy environment if resources are locally plentiful, selection should favour spending on acquiring and digesting them, but if locally scarce, it should favour investing in moving to a new time, space or niche to acquire and digest them. To the contrary, if antagonists are locally plentiful, selection should favour investing in moving to a new time, space or niche to escape them, but if locally scarce, selection should favour spending on defending against them. Science too has antagonists with which it must cope in various ways – pseudoscience, intelligent-design creationism, some extreme animal-rights activism, even some of the "isms" popular in the humanities according to some.

Generally, small organisms are more vulnerable to predation and less so to parasitism while large ones are more vulnerable to parasitism and less so to predation. Populations equally limited by both should be driven towards intermediate sizes. However, populations more limited by predation than parasitism are liable to engage in arms races with their predators (whether individually or with groups of social predators) as to who can become larger thus favouring a larger size, while populations more limited by parasitism should engage in an arms race with their parasites as to who can become smaller thus favouring a smaller size. Similarly in science, smaller labs may be more vulnerable to being absorbed by larger ones while larger ones are more vulnerable to the parasitism of non-performance of some members, even to fraud. What principal investigator with a very large staff can really keep track of what all those people are doing? The fact that labs, research teams, multiple-authored publications and so on are becoming more common (for a discussion see Greene 2007) may indicate that non-performance is fairly well controlled in science although it is the case that fraud, while rare, is believed to be becoming somewhat more common. This does not deny that much of the presumably high performance of large labs currently is being achieved at the expense of those lower down on the totem pole within them.

3.3.6 Individual and demographic growth

In considering the selective effect of density, scale, patchiness and kinds of uncertainty for example, in each case we considered the effects on consumption or individual growth and production or demographic growth separately – *given* low or high density measured in size; or *given* low or high density measured in numbers, etc. But what of individual versus demographic growth? What explains the allocation of effort to these? (For more on this "chicken and egg" problem see Blute 2007.) In prokaryotes like bacteria, growth and division are quite tightly coupled, but in eukaryotes (the more complex kinds of cells present in plants, animals and fungi as well as constituting some unicellular organisms) that is not always the case. For example, a eukaryotic cell has an initial growth phase, a period when its genes are duplicated, a second growth phase, followed by a cell division phase and the two growth phases can be quite variable in amount and duration. Moreover, they may give rise to offspring that are smaller, the same size or larger than themselves (e.g. the latter by growing out in a filament to four times their length, say, and dividing once producing two offspring twice their own size and then again immediately producing four of their own size – conventionally called offspring but in actuality grand-offspring).

Evolutionists commonly say that natural selection acts to maximize "fitness", most often defined as relative success in a population at producing mature offspring. Commonly, it is assumed that "diverting" resources to individual growth is an investment in, and favoured, only in so far as it contributes to, future demographic success – we eat and grow in order to produce offspring. However, it has also not uncommonly been said that, in some groups, particularly branching organisms such as plants and fungi, offspring production looks like growth by other means. In economics, the relationship between households and firms is conceived of as a circular flow in which households provide firms with factors of production such as land, labour and capital, receiving a flow of rents, wages and dividends in return while firms provide households with goods and services, receiving a flow of prices in return. However, it is assumed (and is in fact historically the case) that households antedated firms, implying that ultimately consumption is the end and production a means to that end. This problem of what is means and what is end is apparent in the sociology of science. Hull viewed "curiosity" as a necessary condition for science but thought it otherwise relatively uninteresting (1988:e.g. 341). The important mechanisms of conceptual evolution in science to him were credit (descent) and checking (selection). Considered separately, scientists do research with the goal of maximizing "personal knowledge" but teach and publish papers with the goal of maximizing "public knowledge" (Ziman 1968) and, not incidentally, gaining credit for having done so. But like organisms conceivably growing in order to produce offspring, do scientists do research in order to teach, publish and gain credit? Or, like organisms conceivably producing offspring in order to grow, do scientists teach, publish and gain credit in order that others be enabled to acquire more knowledge? The fact that mammalian species, for example, tend to be roughly arrayed along a continuum of those containing more-numerous smaller members to those containing fewer larger members suggests that the two functions are substitutable for each other. It is strange to think that perhaps in science, those who teach and publish less may actually know more! The simplest assumption and probably the most empirically justified one as well is that of a stable population in which fertility equals mortality so that consumption (decreasing the death rate) and production (increasing the birth rate) are equally beneficial (and on occasion we will make use of that assumption).

Of course, from a gene-centred point of view, none of this variability in sizes and numbers might matter much because individual growth may be viewed simply as demographic growth in situ, so to speak. The circular chromosome in a prokaryotic cell can contain multiple copies of its genome, even an uneven number of such! Ancestrally, it is generally accepted that duplication of eukaryotic cells probably took place with a closed mitosis (intact nuclear membrane), and they may have been multinuclear as are some today, including some of our own white blood cells for example. And of course multicellular organisms contain many cells and eusocial colonies like the ants, bees and wasps contain many individuals. So from that point of view, individual growth may simply be another form of demographic growth and, in eukaryotes particularly, the original form. Similarly, whether individual or social – learning is still learning. Whether acquired in a single scientist or a lineage – knowledge is knowledge.

Turning from the ends–means question (about which more will be said in Chapter 8 on the evolution of complexity) to ecological considerations, individual and demographic growth may be limited by the same ecological factors (e.g. food) or by different ones (e.g. food and available nesting sites) and if by different ones, those may or may not be correlated with each other in various ways. For example considering density, if limited by the same factors or by different ones that are positively correlated, then organisms may simply grow and reproduce repeatedly in concert – an iteroparous life cycle. On the other hand, a negative association could give rise to an semelparous life cycle, usually understood as growth followed by reproduction, but could be the reverse as with those offspring born bigger than their parents who reproduce again before they grow! This alternative, not commonly recognized, could be called "reverse semelparous". And then of course, the phasing of the two activities could be somewhat different so that some organisms neither continue to grow throughout their entire life cycle nor stop growing completely at reproduction but instead the two overlap but only partially.

Undoubtedly the careers of scientists and scholars are similarly variable but a kind of average pattern has recently emerged from data (Brumfiel 2008, and references therein). The youngest scientists and scholars publish fewer papers than the oldest ones. Presumably this is because they are still putting more effort into consuming (learning) than they are into producing (specifically publishing). As productivity increases through the late twenties and the thirties, the impact factor of the journals their articles are published in declines somewhat – quantity of production is being achieved at some cost in quality of production. However, productivity increases again among the oldest, i.e. those in their fifties and sixties, with citations remaining stable. Presumably this is possible because the allocation of effort has been reversed relative to that of the youngest – more total effort is going into producing than into consuming. The role of teaching was not included in the study.

3.4   NOMOTHETIC OR IDIOGRAPHIC?

The study of evolutionary ecology certainly appears, like population genetics or metabolic ecology, to be a nomothetic enterprise. The kinds of statements made above about life histories for example – under these conditions selection should favour this and under those conditions that – are not predicated on historically specific clades, branches of the tree of life, but on universally specified conditions and characteristics scattered across the tree of life. So why then did we begin by suggesting it is a kind of half-way house or hybrid? That is because in the final analysis, the statements that can be verified empirically often end up having to be conditioned taxonomically, even if at a very high level. At the very least, the parameters commonly differ by taxa. Hence for example in the factor analysis of life history characteristics in mammals by Bielby *et al.* (2007) mentioned briefly in Section 3.3 above and described in more detail in the appendix, the data were also analysed controlling for phylogeny (method of phylogenetically independent contrasts – see below). While the same two "timing" and "output" factors emerged (albeit with a somewhat reduced proportion of the variation explained), now the "output" factor explained more variation than the "timing" factor (as it did for the basic analysis performed separately on each of 4 of the 17 orders included for which data on a sufficient number of species was available). Timing is apparently more variable at a higher taxonomic level, whereas output is so at a lower – an apparently historical effect.

Mace and Pagel (1994) proposed that methods developed for controlling for historical effects in biology should be employed in cultural evolutionary research done using comparative methods as well. Relationships that obtain between the ecological environment and the characteristics of societies or among the characteristics of the latter themselves might indeed be causal, but they could also be spurious, attributable instead to phylogeny. For example, was the establishment of early parliamentary democracies attributable to the ecological conditions found in frontier societies with their room for population expansion as found in Canada, Australia, New Zealand and South Africa for example? Perhaps so, but perhaps not. Perhaps instead they were simply attributable to a common historical descent from the British system. After all, there is the contrary example of the United States, also a frontier society, but which established a republic rather than a parliamentary system – inspired in part according to some historians by the enlightenment theorists who also influenced the French revolution. Methods for making phylogenetically independent contrasts are somewhat more complex than simple controls for spuriousness in regression or multiple regression. That is in part because historical effects in evolutionary processes can manifest themselves at many different hierarchical levels. Sometimes the use of these methods confirms the apparently causal nature of associations and sometimes it refutes them. For example, Holden and Mace (2005, and references) confirmed there is a significant negative relationship between pastoralism and matrilineality in 68 Bantu societies, even controlling phylogenetically for cultural descent among them using the tree of Bantu languages – that "the cow is the enemy of matrlineality" as the authors put it. On the other hand Mace and Jordan (2005) found that, in 56 old world societies, phylogenetic controls disconfirmed the idea that there is an association between the sex ratio at birth (high indicating more males) and brideprice as opposed to dowry (although that relationship was confirmed for adult sex ratios for obvious reasons).

In the history, philosophy and sociology of science, views differ radically on "history versus science". At one extreme is the view such as that of Huff (1993) and Drori (2003) that science is, in effect, an historically specific clade. In its origin, two kinds of corporate entities or organizations were particularly important – the self-governing institutions of higher learning which arose in the twelfth and thirteenth centuries (Huff 1993) and of course the scientific societies and their journals which arose in a variety of European cities, mainly in the seventeenth century. From there, science has diffused around the world and, although becoming modified in the process, the kinds of scientific institutions which emerged and with which we are familiar in the west like societies, journals, disciplines and so on are today recognizable everywhere (Drori 2003).

Towards the opposite extreme is the traditional philosophical view that science is a universal mode of thought – whether hypothetico-deductive or whatever, one which, by implication, can and has been practised any time and any place humans are found. Fact: a friend I have agreed to meet for coffee does not show up. Hypotheses: she is just late; she forgot; she had to work late. Deductions: if I wait, she will appear; if I call her at home, she will be there; if I call her at work, she will be there. Test: I wait; I call one place or the other, etc. However, in everyday life we do not normally aspire to employ laws of nature as major premises in such explanations (deductive-nomological) nor do we usually try to explain very diverse groups of facts with the same theory as does science. Beyond the minimum of such a probably universal mode of thought utilizing logic and evidence, to count as science requires its institutionalization dedicated specifically to the acquisition, preservation and dissemination of knowledge. Science is a sociocultural phenomenon as virtually all modern students of the subject like Needham, Merton, Ben-David, Kuhn, Nelson, Hull, Huff and others have maintained (for a review see Huff, 1993: chapter 1). Like other social roles or statuses, it seeks not only to consume (acquire knowledge in this case) and produce (e.g. today publish papers, teach undergraduates) but to replicate in the sense of training successors (today graduate students) who themselves are capable of acquiring and disseminating knowledge. The point, however, is that scientific organizations and institutions do occupy a distinct niche in society – one different from kinship, religious, political, or economic ones for example, albeit at times they can and do establish intimate (symbiotic) relationships with others such as with the state or particular industries. According to the guild-like rather than clade-like view, science flourishes wherever conditions such as adequate support are available. Contrary to the view that science is an institution of solely western origin, some advocates of "multicultural science" (e.g. Teresi 2002) are convinced that scientific institutions of a kind have been present in all societies with intensive agriculture and state-level political organization (sometimes called ancient civilizations).

If we move from the issue of the clade-like or guild-like nature of science in general to its more specific characteristics such as those discussed in Section 3.3 above which should be favoured by selection under universally specified conditions, it is likely that they too are sometimes, but not necessarily always, clade specific. Since findings about the length of publications was mentioned, who for example would not think it likely that while humanists tend to publish more books, and natural scientists more articles, social scientists tend to be somewhat in between? It was recently brought to my attention by an economics informant that accomplished and high-status economists tend to go on to write textbooks and, these days, even some popular books. To the contrary, it is common in many other disciplines like sociology for textbook writing to be a relatively low-status activity which brings little credit. There, textbooks tend to be looked on as commercial ventures. But as my informant pointed out, it might make a certain kind of sense for economists to be inclined to give more credit for commercial ventures! However, I suspect that the real reason is an historical tradition in that discipline going back to Paul Samuelson and before him to Alfred Marshall. There are undoubtedly many other historical particularities associated with the "reward system" in different disciplines and even particular fields and topics within them.

3.5 HISTORY AND NECESSITY

In evolutionary biology at least everyone acknowledges the legitimacy of both the historical "tree-constructing" research programme which can be used to answer the question of where something came from and the more conventionally scientific "selectionist" evolutionary ecology programme which can be used to answer the question of why something evolved (Blute 1997). To be sure, when push comes to shove, the practitioners of each tend to emphasize their own approach.

Systematists, for example, not uncommonly call convergent characters "false" similarities! They are not false at all of course, they just have a different, non-historical explanation. They have been brought about convergently by selection. Systematics commonly operates on the not really satisfying assumption that the best explanation for something is the one that requires the fewest changes to have been brought about by selection – history is always the first factor extracted so to speak. In fact, even if similar numbers of changes are required, systematics will give priority to history in explaining similarities. For example all other things being equal, a sequence of eight bases which differ in related groups, with four similar and four different, e.g. ATTG/CTAG and ATTG/TCGA, will be interpreted as having had a common ancestor with ATTG in the first four positions which diverged in the other four, rather than as having had different ancestors reflected in the second four which converged in the first four.

Selectionists, on the other hand, like to emphasize the amazing changes that selection can bring about, particularly given enough time. For example, as a textbook case Danchin, Giraldeau and Cézilly (2008: section 5.3) illustrate how in various Salamander taxa, from metamorphic ancestors (i.e. with distinct acquatic larval and terrestrial adult forms), some have lost the adult form (becoming sexually mature as larva – a phenomenon called paedamorphism), some have lost the juvenile form (becoming direct developers to the adult form), some have even reacquired much of what they lost again, and some have evolved a flexible form which develops differently depending upon conditions! Of course, such an amazing variety of changes could not be known to have taken place without a phylogeny to map them onto. Selectionists emphasize not only the amazing changes that can be wrought but how more or less the same changes can be wrought in different lineages. Multicellularity, for example, has emerged a number of times independently in different lineages. In the introductory chapter we mentioned Simon Conway-Morris' (2003) argument that if the tape of life were rerun, things would end up pretty much the same, including with something very much like ourselves. A follow-up collection of essays by a variety of authors included one by McGhee (2008) which raised the possibility of something like a periodic table of life. McGhee shows that not only similar changes, but similar sequences of changes in the same order, can be brought about by selection in different lineages. Hence, for example, in the locomotion of both invertebrates and reptiles from legless crawling, a sequence of walking with legs, swimming in 3-d with a fusiform body, and flying with wings have emerged, i.e. branched off in succession. In other groups earlier but not later (amphibians – crawling, walking and swimming) or later but not earlier (mammals – walking swimming, flying) parts of the sequence are present. Similarly, cultural ecologists and archaeologists are confident that horticulture using hand tools and intensive agriculture using draft animals and ploughs emerged successively from hunting and gathering independently in a number of societies (Blute 2008b) as did rank societies and states from early segmentary societies (J. Marcus 2008).

Those who apply population genetics-style models to cultural evolution such as Boyd and Richerson for example, tend to defend a purely selectionist approach to history (1992). Even while controlling for phylogeny, in a seemingly contradictory fashion, a selection first and only perspective can be advocated. For example, in the introduction to Part II of Mace, Holden and Shennan (2005) which moved on from studies of whether cultural evolution is tree-like to studies employing phylogenetically independent contrasts, Mace defends the need for the latter, not as the need to control for phylogenetic inertia, but as necessary simply because two traits being correlated may be an artifact "for reasons other than the two being functionally linked" (2005: 202). On this view "phylogenetic inertia" is misnamed. Continuity in evolution is a result of past evolution and subsequently stabilizing or purifying selection. Selection is not just responsible for change, although that is commonly emphasized, but it is also responsible for maintaining stability. But if that is the case, why not, in traditional nomothetic style, control for those "other reasons" directly? Presumably this is because either no hypotheses or no data is available about what those other reasons might be. It remains the case, however, that what is in fact controlled for in these studies is taxonomic membership, i.e. those historically specific things.

Hence while the legitimacy of both research programmes is widely accepted in biology, it would be misleading to deny that tensions do not still exist. One of the great tasks for the future is a more complete integration of the two – controlling for historical causes of similarity in testing ecological principles which is well underway, but also controlling for ecological causes of similarity in historical tree building. The latter has been hindered by the fact that general principles of evolutionary ecology such as those discussed in Section 3.3 above are only now beginning to emerge. In the social sciences, the gap is even wider. While there was a tremendous revival of historical sociology in the last quarter or so of the twentieth century, not only is the explicit use of shared, derived characters uncommon in attempting to answer historical questions, as emphasized in the previous chapter, but there also remains a great gap between historical and more conventional "structural" causal accounts – a gap which Charles Tilly, for example, for long decried. In economics this gap is a large part of what divides the neo-classicists who emphasize contemporary "beliefs, preferences and constraints" who "forgot history" (Hodgson 2001) from traditional institutionalists (but see also Hodgson 2004, and Chapter 6 on agency). The "new institutionalism" in economics but also in a variety of social sciences, has sought to broaden explanations spatially so to speak, moving away from theories of individual action (rational choice, learning, etc., discussed in Chapter 6) and to move more towards emphasizing the causal importance of institutions and their interactions. In doing so, however, it has also tended to abandon traditional institutionalism's emphasis on the importance of history. In that sense, the new institutionalism resembles more neo-classicism in economics than traditional institutionalism. In fact, the problem of "history versus necessity" is imbedded deeply in the practice and history of all of the social sciences. Are differences in the education, occupational achievement and socioeconomic status by race, sex and class a result of a history of slavery, misogyny and oppression which has become embodied in the people and their families themselves or is it a result of current discrimination? It is what divided the two greatest sociologists of nineteenth-century France – Emile Durkheim and Gabriel Tardé, and sociology from anthropology.

Emile Durkheim (1858–1917) is usually considered the founder of sociology in an institutional sense – first course, first chair, co-founder of the first journal and so on and remains rightfully renowned even in introductory sociology courses and texts to this day. In his time, however, Gabriel Tardé (1843–1904) was equally renowned. He won out over the philosopher Henri Bergson for the chair of moral philosophy at the Collège de France and there is not only a "Rue Gabriel Tardé" in Sarlat but also a statue outside the Palace of Justice – a memorial to one of its most famous sons. Durkheim and Tardé debated with each other both in person and in print. They agreed that a new science of the social was required but disagreed most fundamentally over what its basic subject matter should be. Tardé was convinced that the basic subject matter of sociology should be "imitation" (or what has since been called social learning, contagion, diffusion, collective behaviour, socialization or culture in various literatures) – the "inter-mental", while the "intra-mental" was the subject matter of psychology (e.g. Tardé 1903). (Tardé had been a judge and magistrate in his earlier career and he felt that he learned from those many years of experience that criminals learn to be criminals from other criminals.) There were really two Durkheims. There was the "society as a developing organism" Durkheim of *The Division of Labor in Society* and *The Elementary Forms of the Religious Life*. Then there was the structuralist Durkheim of *The Rules of Sociological Method* and *On Suicide*. To the latter Durkheim, the basic subject matter of sociology was to be social "facts", the social "forces" which push and pull people around by exercising "constraints" on them. These provide the "efficient" causes celebrated in the *Rules* (1895,1982) as opposed to the functional, which is also nice to know, and the genetic, in the nineteenth-century sense of historical which, according to Durkheim, is bunk. (The latter position was a strange one indeed for Durkheim to take, given that a form of mythical history at least lay behind his other organic accounts – once upon a time there was no division of labour, no religion, etc.) In any event, he made his structural case well empirically two years later in *On Suicide* (1897, 2006) where he showed that suicidal behaviour is more predictable than death. It is governed by the social forces that act on occupants of different kinds of statuses particularly – the altruistic (e.g. in the military, volunteers, officers, longer in service), the egoistic (e.g. Protestants, divorced, the educated except among Jews), and the anomic (e.g. in times of social change whether for better or worse). So who was right? It seems obvious in retrospect that both were. As described in the previous chapter, in the twentieth century, David Phillips and his associates well documented the importance of diffusion with Durkheim's most celebrated subject matter, no less, suicide. At the same time, it is obvious that sociocultural phenomena in general, let alone suicide, do not spread indiscriminately. Most people who are exposed to publicity about suicides do not therefore commit suicide. Structurally acting forces act selectively, facilitating or hindering, so that something spreads or not, and does so into some niches rather than others such as those documented for suicide in the nineteenth century by Durkheim for example.

On a larger scale it is also part of what tended to divide anthropologists and sociologists. Of course a major division there was that while sociology had it roots in the study of the industrializing nations of western Europe, anthropology had its roots in the study of small, isolated and non-literate societies of the Pacific islands, Africa south of the Sahara and the first nations of the Americas. However, in the nineteenth century both groups agreed that a new science of what to be neutral, we will call the "sociocultural" was required, but disagreed on what its basic subject matter should be. After Tyler, anthropologists generally adopted "culture" as their subject matter, the way of life of a people including languages and "all those other things that are acquired by man as a member of society" (Tylor 1871, 1958:1), i.e. that are passed on historically by non-genetic means. On a macro scale, their solution more resembled that of Tardé in sociology. Sociologists, on the other hand, more often than not adopted the "social forces" structuralist approach of the one Durkheim, whose work on suicide continues to be held up as a model of sociological explanation. Through time there have of course been departures from these prevailing traditions in both disciplines. The British anthropologist Radcliffe-Brown took the inspiration for his structuralism from Durkheim and the study of culture is popular in sociology today but these have been divergently selected departures from the two disciplines' more mainstream and different historical traditions. (The concept of "social structure" will be examined further in Chapter 8.)

3.6   SUMMARY AND CONCLUSIONS

This discussion began with a description of three approaches to selection. It has been viewed by some, like the historical research programme discussed in the last chapter, as an idiographic enterprise. In the final analysis, the causes of selection are so diverse that every case is unique. At the other extreme are approaches such as those of population genetics and metabolic ecology which seek and claim to have found universal laws of evolution. The contrast is well illustrated by sceptics like Basalla and believers like many physicists, astronomers and cosmologists in the existence of extraterrestrial civilizations.

The third approach of the evolutionary ecology of life histories we described as a hybrid. By combining economic and physical laws, it appears able to make many apparently universal statements about entities scattered across evolutionary trees about life histories for example, whether biological or sociocultural. Examples include those pertaining to what selection should favour under different conditions of density, scale, patchiness and uncertainty which were illustrated with examples from the sociology of science, but they should be applicable to any evolving sociocultural subject matter. On the other hand, not uncommonly, such principles in both the life and social sciences have been found to be limited or conditioned taxonomically, i.e. by historically specific entities. Whether that will turn out always to be the case remains uncertain. Both the idiographic taxonomic and the nomothetic evolutionary ecology research programmes are accepted as legitimate in biology although tensions remain. In the social sciences the gap is wider – historically it was a large part of what divided institutional from neo-classical economics, Tardé on imitation from Durkheim on the social facts or forces that constrain individuals as the basic subject matter of sociology, and anthropologists on culture from sociologists on social structure as their basic subject matters.

It is sometimes said that the life and social sciences are no different from the physical sciences because the latter also have their historical side – in geology and cosmology for example. The comparison is a false one, however, I believe because those who champion the historical perspective in the life and social sciences see history as coming first and foremost with causal principles as secondary. In the physical sciences the reverse is normally the case. Hence for example there are general principles known about how islands form – by volcanic, sedementary or coral means, and while an investigator might have an interest in, and do research to tell the story of, the history of some particular island or group of them, that is always understood as the story of the working out of those general principles in a particular case. A more apt comparison would be with the "multiverse" or "multiple universes" theory (e.g. Smolin 1997) – the theory that the big bang gave rise to or continues to give rise to not one, but an infinitude of universes – each with its own unique constants and even laws of physics (for an accessible introduction see Folger 2008). While some distance from the mainstream, the theory nevertheless has moved in the past couple of decades from science fiction into serious cosmology and physics – the latest phase of the Copernican revolution according to some, each stage of which has seen an expansion of the known universe. If it were to turn out to be the case (and it remains unclear whether we could ever find out if it were), then we could argue that our physical universe too is fundamentally an historical phenomenon in the sense commonly used in the life and social sciences. Obviously a significant part of the future of both the life and the social sciences lies in a greater integration of the two research programmes – controlling for historical causes of similarity in testing ecological and structural principles and controlling for ecological and structural causes of similarity in historical research. It may turn out that there is a greater opportunity in the life and social sciences than there is in the physical sciences for actually answering the question as to whether general principles are always ultimately historically limited or whether, instead, historical phenomena and events can always ultimately be explained by general principles. In the meantime, the only honest answer any life or social scientist can give to the question "history or necessity?" is "history *and* necessity."

3.7 APPENDIX ON R AND K SELECTION THEORY

One of the most general theories of evolutionary ecology ever proposed was that of density-independent or r versus density-dependent or K selection (MacArthur 1962, MacArthur and Wilson 1967, Pianka 1970). It was based on the S-shaped logistic function used to describe population growth as well the diffusion of innovations and much else besides.

$$\frac{dN_t}{dt} = r\frac{(1 - N_t)N_t}{K}$$

The equation describes population growth which is "density-dependent", i.e. where factors limiting growth said to be resource depletion and environmental degradation increase as a linear function of population size itself. It relates the rate of growth of a population at time t (the tangent to the curve relating population size to time) to the existing population size (N_t) and two parameters – the intrinsic rate of increase (r) and the ceiling or carrying capacity of the environment (K). Initially, logistic growth approximates exponential growth because the expression in the brackets approximates 1 and hence dN_t/dt approximates rN_t, the expression for exponential growth. The maximum growth rate (tangent to the curve) is at $K/2$ and the growth rate declines thereafter symmetrically with its previous increase until it reaches 0 at $N = K$ when population size itself levels off.

The theory was concerned with predicting life-history characteristics under different density conditions. According to the most all-encompassing version of the theory at low population densities (or with a history of catastrophes) relative to resources, selection favours maximizing r by rapid development of a small body, an early age at first reproduction, numerous small offspring in a batch, semelparity or at least short inter-birth intervals, low levels of parental care and a short life cycle – a "productivity" or "quantity" strategy. The idea behind this seems intuitively plausible. If both environmental resources for oneself and places for offspring are plentiful, then the following should obtain. (i) Somatically, one should devote oneself more to "productivity" in acquiring environmental resources and less to "efficiency" or to engaging in social conflict over them. (ii) One should devote oneself more to reproductive functions (assumed to be the objective of life) and less to somatic functions. (iii) Reproductively, one should devote oneself more to the quantity of offspring rather than to their quality. On the other hand, at high densities (or with a history of bonanzas) relative to resources, selection should favour coping with K, the ceiling or carrying capacity of the environment by slow development of a large body, late age at first reproduction, few large offspring in a batch, iteroparity with long inter-birth intervals, high levels of parental care and a long life cycle – an "efficiency" or "quality" strategy. There have been a number of applications of the concepts of r versus K selection to sociocultural subject matters such as Hull's (1978b, 1988) on the strategies of scientists and Blute's (1982) on the product life cycle in marketing theory. Perhaps the most extensively developed was Fog's (1997, 1999) on "regal" versus "kalyptic" societies (resembling Spencer's distinction between militant and industrial societies respectively) although Fog ultimately divorced his distinction from the evolutionary–ecological one, presumably because his r and K were more or less the reverse of the evolutionary ecology one.

The theory was enormously influential in evolutionary including behavioural ecology (Danchin, Giraldeau and Cézilly 2008:139) and became a textbook standard. However, it gradually declined somewhat in significance in the technical literature for a variety of reasons. Most derivations were intuitive rather than formal, and some not so intuitive at that. Distinguishing density based on sizes and numbers separately while interpreting productivity as feeding versus efficiency as digestion for the former, and quantity of offspring versus deriving more grand-offspring from fewer offspring for the latter as we did in Section 3.3 above does result in the kinds of predictions about size adaptations in consumption and numbers versus size adaptations in production that r versus K selection theory expected. However, the inclusion of socially antagonistic and cooperative strategies also expected at high densities (discussed in the next chapter) complicates even those expectations. Moreover, as we saw, the logic of this analysis does not necessarily lead to the association r and K selection theory expected between consumption and production strategies which should instead depend upon whether ecological factors controlling sizes and numbers are the same or different, and how, if different, they are or are not associated with each other. Of course the main reason why r and K selection theory expected the low density selected to be small, with a short life cycle and with many small offspring was because of the assumption made about offspring production being the objective of life which, as discussed in Section 3.3 above, is open to question at least for some groups. Many other questions arise. Are spatial properties really just a side-effect of temporal ones, size of life span? Some of its intuitive derivations turned out to be not so intuitive at all.

Secondly, when formal derivations were developed, no one suc-
ceeded in deriving such a large number of predictions from the logistic
or any other single equation (but see Witting 2008), particularly if more
complicated demographic assumptions such as age-structured populations or more details of the life history were added. Thirdly, while
empirically some clusters of characteristics do tend to be correlated
in the way predicted – broadly speaking the world does after all seem to
be made up of species ranging across a continuum from those with
many, smaller shorter-lived members to those with fewer, larger
longer-lived members – empirically there appears to be no one single
life-history continuum and no one single ecological cause of it. For
example in a factor analysis using data from 267 mammalian species
from 17 orders and 59 families, Bielby *et al.* (2007) found that it took at
least two size-independent factors – one cluster relating to the "timing
of reproductive bouts" and the second relating to "reproductive output" to explain 70% to 85% of the variance. The first factor distinguished
those species which independently of size (measured as adult female
body mass) mature quickly, give birth frequently and wean their offspring early from the opposite. The second factor distinguished those
that, again independently of size, give birth to large litters of small
offspring after a short gestation period from the opposite. Moreover,
some clusters of characters which do exist tend to be taxonomically
limited, or as a minimum their parameters are different in some taxa
than others.

In discussing some principles of evolutionary ecology including
in science in Section 3.3 above therefore, while density was one variable considered, others were included. The rationale for these is that
ultimately we are unlikely to be able to explain the spatio-temporal
characteristics of organisms without taking explicitly into account the
spatio-temporal properties of the features of their ecological environment which they have evolved in adaptation to such as those of their
resources and antagonists.

4

Competition, conflict and cooperation: why and how do they interact socially?

4.1 COMPETITION, CONFLICT AND COOPERATION

In the social sciences, historically, economists have emphasized competition over social (strategic) interaction although that has begun to change with game theory, an increasingly important part of microeconomics. Anthropologists, sociologists and political scientists on the other hand have emphasized social cooperation and conflict, traditionally in the form of functionalism and a variety of descendants of Marxism, even though common sense dictates that not all social relationships are cooperative nor are all antagonistic. There, too, game theory is becoming more prominent (Abell 2000). Instead of either the competition or the conflict versus cooperation approach to social relationships and interaction, this chapter adopts the biologists' three-cornered distinction among all three. This helps make clear some general principles under which these alternatives should be expected to be observed as well as their likely consequences.

Some examples of the inferences that can be drawn are that things that are often seen as going hand in hand, such as environmental depletion and degradation, can be distinguished as consequences of different courses of action. On the other hand, some things which are traditionally seen as opposed, such as liberal and conservative views of crime as well as conflict and cooperation, may proceed hand in hand. We then go on to explore the use of such concepts in understanding the nature of proto-gender and gender differences and relationships – the most fundamental social relationship that exists among unrelated peers. Many social scientists may be surprised to learn that, far from being conservative, the most widespread evolutionary understanding of these differences and their relationship is more akin to a radical feminist than it is to a conservative (or a liberal) view although other possibilities exist, in particular, combinations of conservative and radical views. The social sciences can benefit from evolutionary theory by adopting their fundamental three-cornered distinction. However, with concepts such as the advantages of specialization particularly under crowding, of portfolio diversification, of economies of scale, of the existence of transaction costs and the economics of conflict – the former have contributed very significantly to how they can be used from there. We will see how the complexity of social relationships can be better appreciated by coming to understand the role of conflict in a cooperative context and that of cooperation in an antagonistic context.

Finding the conditions favourable for the evolution of cooperation is among the most interdisciplinary endeavours underway today – engaging biologists, psychologists and a variety of kinds of social scientists with each other's theories and findings. Among the most promising may be the concept of byproduct mutualism but higher levels of selection are also part of the story.

4.2   DENSITY/FREQUENCY-DEPENDENCE OF THE ECOLOGICAL: TRADE-OFF BETWEEN DEPLETION AND DEGRADATION

In the previous chapter we discussed density-dependence of the ecological – low density in sizes/numbers favouring spending on acquiring/producing a lot and wasting a lot (e.g. primary research/ teaching in science) and high density favouring investing in digesting/re-producing less but wasting less (e.g. secondary research/teaching in science). However, ecological strategies have consequences as well as causes. Low densities favour spending on acquiring/producing which raise densities while depleting environmental resources but also give environmental degradation an opportunity to be restored. High densities in turn favour investing in digesting/re-producing which lower densities while degrading the environment but also give environmental resources an opportunity to recover. It is obvious why spending on acquiring/producing strategies depletes resources but why should investing in digesting/re-producing degrade the environment? It is because the more "good stuff" extracted/"good offspring" produced, the more concentrated too will be the "bad stuff" left or excreted/the "bad offspring" cast off in brood reduction (the latter is common, even obligate in many species). A simple equation for density dependence is (1) below, where P is the frequency of acquiring/producing at cost C and $(1 - P)$ is the frequency of digesting/ re-producing at cost $(1 - C)$. These strategy frequencies and costs are divided by their benefits (K representing the carrying capacity of the environment and N_t representing the population members size/ numbers at time t so that $1 - N_t/K$ represents the proportion of resources which remain available in the environment and N_t/K represents the proportion which have been absorbed into the population – all from the logistic equation in the appendix to Chapter 3). The two expressions representing frequencies times costs divided by benefits are set equal at equilibrium:

$$\frac{PC}{1 - N_t/K} = \frac{(1 - P)(1 - C)}{N_t/K} \tag{1}$$

so that rearranging, the ratio, PC over $(1 - P)(1 - C)$ is equal to $1 - N_t/K$ over N_t/K, i.e. the ratio of resources remaining available in the ecological environment to those already absorbed into the population. At intermediate densities for example, if costs are equal, $P = (1 - P)$ so that acquiring/producing and digesting/re-producing are equally frequent. With resources depleted by 25%, the ratio is 3:1, for depletion by 33 1/3% it is 2:1 and so on.

Because acquiring/producing deplete, while digesting/re-producing degrade, it is not surprising that sometimes populations evolve under the direct influence of what the majority of other members are doing. If some individuals (or all individuals some of the time depending upon perceptions of what others are doing) compete to acquire/ produce, it will commonly pay others instead to compete to digest/ re-produce and vice versa. It can pay in other words to do the opposite of what the majority are doing – a phenomenon called "negative frequency-dependent selection." Negative frequency-dependence will obtain if individuals doing the same thing "interfere with each other in any way" as it is commonly put. Frequency-dependence is a form of gaming or strategic interaction rather than optimizing in the narrow sense as in density-dependence. Equation (2) below presents a simple model of frequency-dependence in which the benefits of each strategy are proportional to the frequency and costs of the other and again they are set equal at equilibrium so that:

$$\frac{PC}{(1 - P)(1 - C)} = \frac{(1 - P)(1 - C)}{PC}. \tag{2}$$

Solving, the total expenditure on each strategy, i.e. PC and $(1 - P)(1 - C)$, is equal at equilibrium. If costs are equal, then the frequencies P and

$(1 - P)$ are equal at equilibrium. Only if costs are unequal, will frequencies be unequal at equilibrium. This tactic of avoiding competition by different specializations is as common socioculturally as it is biologically, and probably accounts for much of the diversification, even fragmentation in science for example.

Density- and frequency-dependence can even be simultaneously operative. For example in science, it might be advantageous to engage in primary research in an uncrowded field *and* when doing so is rare and to engage in secondary research in a crowded field *and* when doing so is rare. For example, density- and frequency-dependent forces may act severally, such as might obtain when two different sources of information (such as genetic loci or places in genomes) are involved which nevertheless effect the same phenotypic characteristic. Under this assumption we can combine equations (1) and (2) from above and, solving them together, the result is that the only equilibrium, i.e. where $PC = (1 - P)(1 - C)$ is at $1 - N_t/K = N_t/K$, i.e. at an intermediate density. On the other hand, if frequency and density act jointly such as might obtain with a single source of information (such as a single genetic locus or place in a genome) is involved, then its joint dependence on density and frequency can be expressed as in equation (3):

$$\frac{PC}{(1 - N_t/K)(1 - P)(1 - C)} = \frac{(1 - P)(1 - C)}{(N_t/K)PC} \tag{3}$$

with the result that the population is in equilibrium at intermediate density but can also be in equilibrium elsewhere – with resources depleted by 20% or 80% with PC and $(1 - P)(1 - C)$ in ratios of 2:1 or 1:2 respectively, for example.

It is interesting that in the previous chapter we were typically talking about four alternatives – consumption versus production and, within each, two alternatives – whether different densities, scales, local densities or kinds of uncertainty. The phenomenon of negative frequency-dependence among them could conceivably be responsible for the preponderance of "quarter" values in scaling relationships. (A simple expansion of equation (3) can show that, at equal costs with pure frequency-dependence, all four alternatives should be present at equal frequencies). On the other hand, as we saw above, values based on "thirds" are certainly possible with pure density-dependence and as derived above, from a combination of density- and frequency-dependence. As we noted in the previous chapter, some of the facts about these values remain in dispute.

In both popular environmental and technical literature it is common to talk of "resource depletion and environmental degradation" in one breath as if they normally go together, particularly at high densities, but that may be in error. That there may commonly be a *trade-off* between depletion and degradation as described in this section so that while some strategies deplete, their alternatives degrade and vice versa is a sobering thought for our current environmental problems. Details matter of course. However, very generally, while currently there is rightfully much pressure to consume and produce less and instead to be more efficient in our use of environmental resources and production processes in order to reduce rates of depletion, a side effect of such shifts unfortunately is liable to be an increase in environmental degradation. Recent human agriculture is an example. In the past 40 years, while land use has increased by only 12%, global crop production has more than doubled – as a result of intensification by means of high yield seed, chemical fertilizers, irrigation and mechanization. The side-effect of this way of obtaining more food, however, has been extensive environmental degradation including of water, land and air quality (Foley *et al.* 2005, 2007). Another example of the downside of efficiency is the environmental toxic effects of recycling e-waste in India and China which have been apparent for some years (Greenpeace International 2005).

It is also worth noting that the same or similar formalisms as employed in this section for selection based on density-dependence, frequency-dependence or their combined effects can also be applied to the other alternatives considered in the previous chapter such as scale or local density. For example, with scale dependence, r/d consumption/production strategies deplete small-scale resources but may allow large-scale ones to recover, while t/s consumption/production strategies deplete large-scale resources but may allow small-scale ones to recover. In some circumstances they can even be applicable to consumption (including acquisition and digestion) versus production (including production and re-production) themselves where the resources required for the two functions such as food for the former and nest sites for the latter perhaps are negatively correlated. In such circumstances, densities low in size but high in numbers would favour consumption strategies and the reverse production strategies. Such a trade-off would be within individuals such as some birds and sea mammals which feed at sea but give birth on rocky shores, but can be between individuals if roles commonly change over the life course as in colonially nesting species. It is also clear that such a difference may commonly divide species which, after all, are largely arrayed along a range from those containing fewer, larger individuals to those containing more numerous, smaller individuals.

4.3   DENSITY/FREQUENCY-DEPENDENCE OF THE ECOLOGICAL VERSUS THE SOCIAL AS WELL AS WITHIN THE SOCIAL: THE AFFINITY OF CONFLICT AND COOPERATION

If density- and/or frequency-dependence can be used as in the previous section to distinguish what are commonly thought to go together (depletion and degradation), they can also be used as shown in this section to bring together what are commonly thought to be necessarily distinct – liberal and conservative views of crime and conflict and cooperation. Density- and/or negative frequency-dependence may obtain not only with respect to various ecologically oriented strategies but also with respect to the ecological versus the social (whether conflict or cooperation). At low densities, given that more resources remain available in the ecological environment, selection will favour competing for them ecologically. At high densities, given that more resources have been absorbed into the population, selection will favour focusing effort on social interactions, whether conflict or cooperation. We will begin first with social conflict, and then proceed to consider social cooperation and their combination.

4.3.1   Social conflict

The low-density ecologically oriented strategies have been variously termed passive, indirect, scramble (or, confusingly, exploitative) competition, while the high-density social conflict strategies have been variously termed active, direct, contest or interference competition. The key distinction is that with conflict as opposed to competition – contact and aggression are involved. At low densities where monkeys are small and there are lots of bananas, the monkeys should compete ecologically, acquiring their bananas from the ecological environment. But at high densities, where monkeys have become large and there are few bananas, the monkeys should employ contact and aggression to acquire their bananas from other monkeys – called "kleptoparasitism" (food theft). If other monkeys have not only acquired the bananas but have eaten them, then selection might favour eating them – cannibalism. If they have not only acquired and eaten them but also turned them into offspring so that monkeys are numerous, then selection might favour producing offspring to parasitize the offspring of others – intraspecific nest parasitism (tricking competitors into raising your young – well known in many bird species, for example).

Such contest or interference competition is common in human economic and political markets, in negative or attack ads for example. What evolutionary theory makes clear is the conditions under which it should be favoured. If most customers or voters are not yet committed (often early after the introduction of a new product class or early in a political campaign) then selection should favour simple competition. Ignore your opponents and simply convince your audience how good your product or candidate and platform are. Once more than half of the resources have been absorbed into the population, however, i.e. the majority of consumers or voters are committed, then it should pay to engage in contests – you need to take customers or voters away from other firms or candidates. Overt conflict is not unknown in science as well (see below).

Frequency-dependence between the ecological and social conflict is also a possibility. If some individuals (or all individuals some of the time depending upon what others are doing) compete for some component of fitness ecologically, all other things being equal, it pays others to exploit them socially. And it works in both directions – just as exploiting is favoured if competing is common, competing is favoured if exploiting is common. Put simply, exploitation is favoured if victims are available and victims may have no alternative other than to put up with (or compensate for) being exploited. They do (whatever) for their exploiters as well as themselves. As with frequency-dependence among purely ecological strategies, the total expenditure on the two alternatives are expected to be equal in a population at equilibrium.

Note how exploiters should match their life-history strategies to that of their victims. If victims are consuming, one should consume them. If they are producing, one should produce to exploit their production and so on. Consider a predatory species. If their prey is adapted for sprinting, then they had better be able to sprint too – be an ambush rather than a pursuit predator. On the other hand, if their prey is adapted for endurance over longer distances, then they had better have endurance too – be a pursuit rather than an ambush predator. Of course many variants are possible – a short-lived parasite might match the life history of a long-lived host with many generations of its life cycle and a small predator might match the size of a large prey by joining forces with others of its kind in social predation – but still, the general principle of matching in antagonistic relationships should obtain.

Such evolutionary "producer–scrounger" games were first described in the 1970s and 1980s (Brockmann and Dawkins 1979,
Brockmann, Grafen and Dawkins 1979, Barnard and Sibly 1981,
Barnard 1984a) and have been applied to a whole host of biological
phenomena since including: food theft; theft of other resources such as
nests or burrows; cannibalism; gender interactions involving force and
manipulation; strategies of mate competition among members of the
same gender which, rather than simply out-racing or out-enduring
rivals employ force, stealth or fraud (e.g. fighting, sneaker and trans-
vestite males respectively); and brood parasitism. In the human social
sciences, they have been proposed to be applicable to expropriative
crime (Cohen and Machalek 1988).

Of course counter-exploitative strategies may evolve (Barnard
1984b). The exploited may evolve to evade in anticipation of attacks,
defend at the time of, or counter-attack subsequently. However, in the
absence of historical obstacles, the exploiters would then be favoured
either to exploit those in turn (some thieves steal the armoured car as
well as the money), or alternatively, to retreat to competing ecologically themselves (doing whatever for themselves). The overall budget in
the sense that the total expenditure in a population (frequency × cost)
devoted ecologically and to social conflict (whether between individuals in the former case or within them in the latter) would not thereby
be altered, i.e. would be equal at equilibrium.

Density-dependence of ecological versus socially exploitative
strategies may be viewed as a formalization of the traditional liberal
view that crime is a consequence of deprivation, of a lack of opportunities for acquiring resources honestly. Frequency-dependence of
ecological versus socially exploitative strategies may be viewed as a
formalization of the traditional conservative view that crime will
always be with us. It is interesting therefore that in evolutionary
models these two polarized views can be combined. And it is particularly interesting that with the one "joint effects" model of combined
density- and frequency-dependence as in equation (3) in the previous
section, a society can be in equilibrium at lower (1:2), intermediate (1:1)
or higher (2:1) crime levels.

4.3.2 Social cooperation

We turn now to social cooperation. Density- and/or frequency-
dependence may obtain between ecological competition and social
cooperation as well as social conflict. Whether because of high density

or because of a high frequency of the ecologically oriented, the availability of potential victims also means the availability of potential partners. Social cooperation can take two forms, one of which is economies of scale in which the role of the partners is not different, i.e. like antagonists, they match. If individuals working together can achieve more per capita than they can working separately, social cooperation based on economies of scale can be an alternative to social conflict – acquiring partners rather than victims. Among animals, this is commonly observed in social predation, crowding together for heat conservation, travelling in flocks or schools for reduced per-capita resistance, and living in groups because innovations discovered by individuals can be learned socially from each other. Social cooperation based on economies of scale tends to be more common in defence than in resource acquisition because of the advantages of grouping on perimeter reduction, as is indeed the case with humans with their walled cities, travelling in caravans, etc.

More common than cooperation based on economies of scale, however, is that based on a division of labour within, or specialization and exchange, i.e. trade between, individual entities. We have known since Adam Smith's (1776, 1977) wonderful description of a pin factory in the eighteenth century the very great advantages to be reaped from a division of labour within firms, and similarly, of specialization and exchange, i.e. of trade between them (hereafter for convenience, both will be referred to as specialization). And since Emile Durkheim (1893, 1964), in the nineteenth century, we have been aware of the association of specialization with crowding as well. Whether because of some intrinsic (i.e. historic) comparative advantage or because of economies of scale, specialists are normally more efficient, and commonly much more efficient in the range of a niche in which they specialize than are generalists in that range (which is not necessarily to deny the success that generalists can have overall, only when faced with a team of specialists). With density dependence, crowding should push a generalist ecologically oriented population towards becoming a population of diverse socially cooperating specialists, whether within or between individuals. Abandoning half a niche would be favoured by selection if one were slightly more than twice as efficient in the remaining half for example and once one kind of specialist emerged, another or even others likely would as well. Note to be favoured by selection, while returns to each kind of specialist would have to exceed those to generalists, they would not necessarily have to accrue equally to each kind. One or another could be forced into occupying a smaller segment of the niche – "making the best of a bad job" as it is sometimes called, resulting in a biased allocation to the two groups. Again, frequency-dependence is also possible. A low frequency of ecological generalists should favour a mix of cooperating specialists and vice versa.

4.3.3 Both

In summary on social cooperation, cooperation is generally favoured under the *same* conditions that social conflict is – at high densities or when rare. It is not surprising then that they tend to go together – conflict in a cooperative context or cooperation in an antagonistic context. After all, even Marx saw cooperation within classes in the context of conflict between them and Lenski (1984), after Sumner, liked to speak of "antagonistic cooperation". Choi and Bowles (2007) and Bowles (2008) for example have termed this "parochial altruism" – arguing that the other side of conflict between groups of animals and humans including bands of hunter-gatherers, early tribal societies, ethno-linguistic groups and even modern nations is cooperation within them, or, put differently, that the other side of cooperation within them is conflict between them.

It is worth noting that in whatever form, social conflict and cooperation should have opposite effects on the population as a whole. Social conflict is likely to have a negative effect on the population as a whole. This is essentially because intraspecific resource transfers – food theft, cannibalism, intraspecific nest parasitism, etc., consistent with the well-known inefficiency of transfers between species – are presumably made with far less than perfect efficiency. Cooperation, on the other hand, at least initially, should allow growth to continue to accelerate rather than to begin levelling off, but as a consequence, the carrying capacity of the environment is approached under acceleration rather than deceleration – with potentially catastrophic results.

Both conflict and cooperation are evident in science. There, in favour of Hull (1988) and contrary to Maynard Smith (1988), the overt conflict Hull sometimes observed and reported on in his sociological study of scientists was probably not unique to the particular field, systematics, he chose to study – albeit systematics was an old and probably crowded field at the time relative to its resource base (funding resource base – not species to be classified!). On the other hand, in favour of Maynard Smith and contrary to Hull's "what's good for scientists is good for science" functional interpretation, there is no reason to believe that all forms of competition in science, specifically contests, benefit science as a whole any more than there is to expect that food theft, cannibalism or intraspecific nest parasitism benefit a biological population as a whole or that attack ads in politics benefit democratic institutions.

Finally, density- and/or frequency-dependence can apply to different socially antagonistic/cooperative strategies themselves. If some individuals cooperate/engage in conflict in one way, it can pay others to do so in its opposite. A good example is alternative mating strategies. In some species, males for example come in more than one form – large ones fight for mates and small ones sneak. It is obvious that with density/frequency-dependence possibly obtaining among ecological strategies, between the ecological and the social, and among social strategies, the possibilities are extremely numerous. In Section 4.5, some of the concepts from this and the previous section will be applied to understanding gender differences and relations but, first, some background on genders is necessary.

4.4   GENDERS: SOME BACKGROUND

Social scientists understand the terms "sex" and "gender" differently than do life scientists. To social scientists sex is the singular of "sexes" – indicating the biological component of female–male differences. Gender on the other hand is the singular of "genders" – the socioculturally structured and constructed layer of differences and relationships including inequalities which are added on to, or even distort biological differences and relationships (Mikkola 2008). In short, to social scientists sex is biological while gender is sociocultural.

Unfortunately for this biological versus sociocultural distinction in the social sciences, both the concepts of sex *and* of gender differences and relations have a history of usage in the life sciences to indicate different *biological* phenomena, a history which long antedates this relatively recent usage in the social sciences. To life scientists, sex is about genetic recombination which can and does take place in the absence of the familiar female/male distinction (or female/male functions distinction in hermaphrodites). In the absence of gender or kinds of gender-functioning distinctions, sex is restricted to between different mating types which, while commonly two, can range from two to many. These "isogametic" species including many protists, algae and fungi engage in sex without any of the familiar differences in the sizes or numbers of gametes, let alone the secondary sexual characteristics like the "weapons" and "ornaments" that we associate with gender differences. Between these and "oogametic" groups like we animals in which there is a qualitative difference between extremely large, few wholly immobile eggs, and extremely small, numerous mobile sperm and in which females are *defined* as egg producers and males as sperm producers, lie many "anisogametic" species. In anisogametic groups there is some but much less differentiation of gametes in numbers and sizes (e.g. ratios of 4 or 8 to 1 say rather than orders of magnitude differences) and with variable differences in mobility (the micro only, both or neither may possess flagella which make them mobile). Anisogamy is assumed to have emerged from isogamy, and oogamety (or oogamy) from anisogamy, probably in both cases a number of times. Hence anisogamy is thought of as a kind of "proto" gender difference and hence understanding the selection pressures that created and sustain it is thought to be key to understanding the foundation of the nature of gender differences and relationships.

It is commonly said that the biological difference between genders, particularly in mammals, is that one gender (usually, but not always females) devotes more resources to "parental effort" while the other (usually, but not always males) devotes more resources to "mating effort". Obviously, the "parental effort" of the former, when devoted to sexually produced offspring, is something that contributes to the fitness of both parents but the meaning of the "mating effort" of the latter is ambiguous. Does it mean a suite of naturally selected adaptations by which males for example make sex and the production of sexually produced offspring possible and hence contributes directly to the fitness of both parents? Alternatively, is it a suite of sexually selected adaptations by which males/sperm compete with other males/sperm for females/eggs and hence contributes directly to the fitness of the male parent only, and which indeed can be viewed as parasitic of one or more females since it is effort which could have been devoted parentally? Or is it some proportion of each?

The first interpretation has largely prevailed in the history of biological theories of gender differences and relations. This was certainly true of the classic Parker, Baker and Smith (1972) and Bell (1978) theory of the origin of anisogamy. In an ancestral population in which it was assumed that there was a fixed amount of cytoplasm available to invest in gametes and in which every individual (gamete) could mate (form zygotes) with any other, the formal condition for the evolution of anisogamy proposed was that fitness increased greater than linearly for at least part of the range in the one direction of gamete numbers yielding more offspring, in the other direction of gamete size yielding better surviving offspring, or in both, thus causing the evolution of anisogamy from isogamy by disruptive natural selection. (If returns always increased only linearly, then averaging the benefits of numbers and size, producing an intermediate number of intermediately sized gametes would be just as fit as the extremes.) It was assumed that these accelerating naturally selected returns probably accrued to macroga- mete production because of crowding. Subsequently, microgametes or their producers turned to sexual competition with other microga- metes, evolving to fuse preferentially with the macrogametes thus getting the best of both worlds – more zygotes and better zygote survival. In short, proto-males were, and males remain, reproductive parasites of females, and gender relations are a "producer–scrounger" game (Parker 1984) in the same class as food theft, cannibalism, and intraspecific brood parasitism for example.

This interpretation has also been true of the line of thought about other gender differences in classic articles running from Bateman (1948) on greater variance in male than in female reproductive success through to Trivers (1972) on differences in parental investment to much of contemporary evolutionary psychology. Whichever gender invests more in each offspring is said to be a scarce resource for which the other competes. Usually the former are females because eggs are said to be expensive and sperm cheap but the difference in parental investment extends beyond that particularly in mammals to internal fertilization, gestation, lactation and post-parturitional paren- tal care. This view of males as reproductive parasites is more akin to a radical feminist than to a conservative (or a liberal) view of gender differences, albeit feminist evolutionary biologists, evolutionary ecolo- gists and primatologists have been quite critical of its unidimensional quality (Liesen 2007, and references therein). In fact, my undergraduate students get the problem with the approach fairly easily.

Play the following kind of game with half female and half male students. Females are constrained to invest in say two $50 packages. Males are constrained to invest in say ten $10 packages. The rules of the game are that some kind of investment (buying stock, starting a busi- ness) must be made jointly by a male and a female and they share the profits equally. Set them free to play and you see many of the phenom- enon associated with sexual selection. Males race from female to female trying to get their attention, often making several proposals, and females are choosey both between proposals and males. You over- hear females saying things like, "Now explain that to me again" and

"Wait a minute until I hear what this other guy has to say". The interesting part takes place when you end the game and ask them "How, if you could, would you change the rules of the game?"

They get the problems immediately. The females say (a), "Why do I have to have a partner? It is unfair if I put in $50 and he puts in $10 and we share the profits equally. Why can't I invest on my own?". Indeed, why don't females become parthenogenetic? As Turke (2008:252) remarked recently, the Parker *et al.* (1972) theory "provides little basis for explaining why egg producers (females) do not regularly opt out of this exploitative system by bypassing meiosis to produce diploid gametes that do not require fertilization." They assumed that sex was fixed prior to the origin of anisogamy. When you tell the students that rule is fixed the females come up with (b), "If I have to have a partner, why can't I have a female partner who would therefore contribute equally?". Indeed, why don't females evolve to mate with other females? Parker *et al.* did consider this possibility – macrogametes would also benefit from uniting preferentially with other macrogametes but assumed it was impossible because the proto-eggs were immotile. However, as we noted above, anisogamy is a diversified phenomenon rather than a uniform stage on the way to oogamety – both micro- and macrogametes may be flagellated, for example. If you tell them, "No, proto-females are historically limited in this second way as well" they come up with a third alternative: (c) "Why can't males vary in how much they invest, and I would pick the one who invests the most?". Indeed, it is impossible to imagine that there would not be some variation in the size of the microgametes and hence why don't the macro always choose the larger, causing the them to evolve back to the size favoured by natural selection? At this point, the males join in, approving of this alternative as well. They point out that if females are constrained to invest in two $50 increments while they are constrained to invest in ten $10 increments then, on average, 80% of their investment is destined to be wasted (although occasionally a bright male asks if they can spend the rest on interfering with other males). But in essence, the problem is clear. In order to accept conventional antagonistic theories of primary and secondary gender relations we have to be willing to assume that in many lineages independently, females are historically limited in a whole series of ways – from becoming parthenogenetic, from evolving to mate with other females, and from choosing males for resources – a series that, in total, is implausible. The conclusion seems inevitable that either males are contributing something ecologically that they are not getting credit for, or females are doing something sexually that they are not taking the blame for, or both.

4.5   GENDERS AND GENDER RELATIONS

Let us go back then to some of the difficulties with the traditional theory of the evolution of anisogamy. First is the assumption of starting from a promiscuously mating population of isogametes with polarization coming about only through the micros coming to fuse dissasortively with the macros. In all known cases, however, populations with isogametes are themselves divided up into mating types in which mating is only possible between and not within kinds. Secondly, a number of authors before and since (reviewed and used in Iyer and Roughgarden 2008) have made the point that microgametes *increase the encounter rate* with the macro (recall the discussion in Section 3.3 on the proportionately greater surface to volume ratio of the small relative to the large which could increase sexual contacts as well as with food). Thirdly, in discussing disruptive selection generally (not specifically with respect to the origins of anisogamy, nor two functions, nor accelerating returns) Rueffler *et al.* (2006) pointed out that unless the mean phenotype of a population is sitting right at the point of minimum fitness, rather than resulting in disruption, selection would likely move a population towards the closest pole. However, for one in that neighbourhood only, negative frequency-dependent selection can lead to disruption.

Sexual selection alone cannot explain the existence of micro and macro reproductive cells because micro and macro spores (asexual reproductive cells) in which the micro as well as the macro can develop parthenogenetically exist in various groups (the mobile "zoospores" of some eukaryotic protists resemble nothing so much as sperm). Ecological differences between genders are common but have tended to be studied separately under the rubric of "sexual segregation" (e.g. Ruckstuhl and Neuhaus 2005) and are mostly ignored in the literature on sexual selection. (Just to take a few examples from the most important classic monographs, in 1978 Maynard Smith devoted a little more than a page to these, 184–5; in 1994 Andersson devoted a couple of pages, 15–17; in 2003 Shuster and Wade devoted their second chapter to the ecology of sexual selection but one following the 1977 Emlen and Oring hypothesis that females distribute themselves according to ecology and males according to females; and in 2005 Arnqvist and Rowe reverted to a couple of pages, 26–27, although they devoted their chapter 5 to sexual conflicts over parental care which is an ecological or direct benefit). This general lack of attention to ecological differences between genders is beginning to change, however – in the study of sexual size dimorphism for example (Fairbairn, Blanckenhorn and Székely 2007).

4.5.1 Anisogamy

Consider then an example of the kind of ecological specialization that could be involved in anisogamy. We assume haploids which, if sexual, have a zygotic meiosis. A generalist asexual cell doubles in size and divides once ($1 \times 1 \rightarrow 1 \times 2 \rightarrow 2 \times 1$). We assume an equal allocation to two kinds of sexual specialists, whether because they are at the mid-point in density, because of frequency-dependence, or even because a low density-selected population and a high density-selected population are hybridizing (in which case mating type alleles would not be homologous). In any event, we assume specialists are twice as efficient as generalists in what they specialize in. It seems natural (but see below) to think of proto-males as "quantity" specialists – both consuming more and producing more while proto-females are "quality" specialists – both digesting more and re-producing more. Hence a sexual macrogamete producing cell quadruples in size but does not divide, the cell itself functions as a gamete ($1 \times 1 \rightarrow 1 \times 4$). A sexual microgamete producing a cell of the opposite consumption specialty similarly quadruples in size but it also divides twice producing four microgametes ($1 \times 1 \rightarrow 1 \times 4 \rightarrow 4 \times 1$). Non-random mating takes place in the sexuals between the two different kinds of specialists. It takes four of the micros for one of them to make contact with and fertilize the macro by which time the mass of the former is virtually gone. After fertilization, the macrogamete contribution to the zygote fuels and directs a two division meiosis ($1 \times 4 \rightarrow 4 \times 1$). These would commonly be called four offspring, but in actuality, they are four grand-offspring. The generalist asexual's yield is (2×1) descendants while the sexuals, in approximately the same amount of time, together yield (4×1) descendants. The micros or their producers underwent two divisions *before* sexual union thus providing the surface necessary to make contact and hence are responsible for the production of offspring. The macros or their producers fuelled and directed two divisions *after* sexual union (the two-division meiosis) and hence are responsible for the production of grand-offspring. In effect in the sexuals, the microgamete producer contributes offspring but the macrogamete producer contributes grand-offspring, both to their joint project of giving rise to descendants.

Note four things. First, it is commonly said that with cloning one becomes two while with sex two become one – one way of formulating the famed "two-fold cost of sex". But here the sexuals, because of the advantages of specialization, have made up for the two-fold cost of syngamy. This is manifested by and explains why meiosis is a two-division process. They have not more than made up for the two-fold cost, i.e. they would not be individually favoured over the asexual, because of course they leave only two descendants each, the same as the asexual. However, if the advantage of specialization were a little more than two-fold (manifested for example by one or more post-meiotic mitotic divisions which are common in eukaryotic protists) then they would. Secondly, both kinds of sexuals make an equal, naturally selected contribution to creating descendants. By providing a disproportionate amount of surface area relative to volume or mass with four microgametes, and commonly the motor power in the form of flagella to get there as well, the microproducer makes contact and therefore offspring possible. By providing a disproportionate amount of volume or mass relative to surface area, the macroproducer makes grand-offspring possible. Thirdly, it is notable that all multicellular sexual organisms have retained a single-celled stage in their life cycle and it is curious why this is so. Some multicellular organisms reproduce asexually by means of multicelled propagules so why do none ever have sex that way – recombining multicelled propagules? It would be difficult to argue that it would be too complicated given the complications of what goes on in syngamy and meiosis. An obvious possible answer is that a necessary material exchange going on at the cellular level is what sex may most fundamentally be all about.

Fourthly, asexual specialists, micro- and macrospore rather than gamete producers, could similarly reap the advantages of specialization. However, there can still be an advantage to the sexuals. Because of the existence of the diploid phase (short in haploids with a zygotic meiosis or prolonged in diploids with a gametic meiosis) a new level of selection has been created – sexual families. (This is the case whatever the breeding system, it is just that in species in which one or the other or both genders mate multiply, sexual families overlap in individual membership.) Darwin himself maintained that selection acts on families as well as individuals. Assuming the genomes do not "congeal", i.e. recombination takes place – sexual families include a variety, not a single kind among their offspring. If uncertainty prevails over what life history strategies or combinations of them would be advantageous (and no reliable cues are available which would favour adaptive phenotypic plasticity), sexual families would have a higher geometric mean fitness (the nth root of the product of their fitnesses rather than their sum divided by n) than asexual kin groups, i.e. a reduced risk of extinction. What social scientists know as portfolio theory – investing in an index fund rather than trying to pick stocks or time markets – is known to biologists as "bet-hedging" (Seger and Brockmann 1987).

That sex in the sense of segregation and recombination creates diversity/change within families, populations and species is one of the oldest and most general theories of "why sex?" although it comes in many varieties. That it is part of the story is made obvious by the scattered taxonomic distribution of parthenogenesis and the fact that few higher taxa are exclusively parthenogenetic. These can be accounted for if parthenogenesis is sometimes advantageous to individuals within populations and so pops up from time to time here or there, but subsequently parthenogenetic populations tend to go extinct while sexual populations and species remain in existence and speciate. However, theories of the advantage of recombination as opposed to gender differences and relations go beyond the scope of our discussion here (for some reviews see Otto and Lenormand 2002, Birdsell and Wills 2003, Agrawal 2006, Otto and Gerstein 2006 and Hadany and Comeron 2008). I will only add one comment on the recombination problem. It seems at least conceivable that what is commonly called a "fair meiosis" and viewed as a mechanism for controlling conflict between the parties at least *arose* as an expression of conflict (Hickey and Rose 1988). The parties fight over the spoils of their association then go their own way (recombination then being a side effect of "every man, i.e. genomic sequence, for himself" rather than team play at the meiotic line of scrimmage). That sex is associated with crowding, both intra- and interspecifically (Bell 1982), sheds no light on this as both conflict and cooperation can be favoured under crowding. It would be interesting to know if gene conversion, including balanced gene conversion, is more common in haploids with a zygotic meiosis than in diploids with a gametic one, i.e. the diploid phase takes hold as selection on the higher level discovers the advantage of bet-hedging. Still, it is mystifying if that is the case, why recombination is not restricted to functional units (see Section 5.2 on the "gene" concept). While recombination is recognized as a difficult and unsolved problem, at the same time, if ecological specialization by proto-gender and gender differences and their non-random mating makes up for the famed two-fold cost, it goes a considerable way towards easing that problem.

4.5.2 Isogamy

Is ecological specialization a possible explanation of isogamy rather than anisogamy? Yes it is. Analyse further the quantity strategy of spending on acquiring/producing versus the quality strategy of investing in digestion/re-producing. For example, acquiring may be composed of eating plus breaking down (degradative metabolism), best achieved by being small and large respectively, while digesting may be composed of building up (biosynthetic metabolism) and excreting, best achieved by being large and small respectively. A similar breakdown is possible for production and re-production – resulting in similarly appearing intermediately sized individuals and gametes, even though ecologically specialized. While fine for explaining what might be called "mesoisogamy", additional assumptions about differences in costs or benefits of the component functions would be required to make it yield micro- or macroisogamy. Having made the point that isogamety could be a more subtle form of ecological specialization, it is also possible that it is a mutually cooperative relationship based on economies of scale or one initiated by exploitation. For example, at low or high densities, an ancestral asexual micro (in the former case) or macro (in the latter case) spore producing population could be transformed into a sexual population by invasion, by frequency-dependence, of a matching alternative which mates with the former exploitatively (meiotically driving, say). Extending the game with the former evading, defending or counter-attacking, the latter could be driven to retreat to making a naturally selected ecological contribution. With everything in equilibrium at equal costs, the two would be in equilibrium at equal frequencies and matching, albeit again more likely "meso" rather than micro or macro.

4.5.3 Sexual competition and selection

Leaving aside isogamety, for which the possibilities are multiple, having made the argument that with anisogamy, proto-males are not necessarily parasites, that both proto-genders, genders or kinds of gender-functioning have naturally selected functions that are mutually beneficial because of the advantages of specialization (and on the family level of bet-hedging), that does not mean that conflict does not enter into sexual relationships. We used the mid-point of density- or frequency-dependence, or both with an equilibrium at equal frequencies, or even hybridization to assume that the two kinds are not only in equilibrium but are so at equal benefits and equal costs which may make sense for anisogametic microorganisms in which the signs of sexual selection like weapons and ornaments are at least not obviously visible.

But to turn to oogamety and more familiar gender differences, what if the consumption and production functions of one gender, say females, are more costly than those of males so that the naturally selected equilibrium is with females/eggs at a lower frequency than males/sperm "packets" (the number of sperm required for a single egg to be fertilized; Noë and Hammerstein 1994, 1995)? When it comes to mating, this leads to something close to the traditional view of gender differences and relations which Clutton-Brock (2007) has argued is more or less right. With more males than females/more sperm packets than required for each egg to be fertilized, with such an "operationally" (Clutton-Brock and Parker 1992), i.e. analytically biased, sex/gamete ratio, because of supply and demand, males/sperm are forced to compete sexually with others of the same kind for even a single female/egg. Because their resources (females/eggs) are scarce, this should take the form of a contest rather than a scramble. The contest can be direct with weapons and intimidation displays directed at other males or indirect with ornaments directed at manipulating females (passive female choice). Because effort that males devote sexually is effort that they divert from their naturally selected functions which benefit females as well as males and which were previously in equilibrium between genders, it is both harmful to the population as a whole and sexually antagonistic to females, and females therefore are also favoured to resist being manipulated and to actively choose, in a similarly sexually antagonistic way, males/sperm who devote their resources more naturally than sexually. If we then assume that just as female naturally selected functions are more costly than those of males, that male sexually selected functions are more costly than those of females, the population will be driven towards an overall sex allocation equilibrium at equal frequencies (see Table 2, line (i)). If we began all over again with the opposite cost imbalances and hence with females/eggs at a higher naturally selected frequency, the result is what has traditionally been called "sex-role reversed". Females compete for males/sperm with weapons and ornaments and males actively choose (Table 2, line (ii)). (If we were not to assume that sexual costs within a gender were

Table 2. *Examples of some possible natural and sexually selected components of gender differences*

	Male				Female				
	Natural		Sexual		Natural		Sexual		
	fr	co	fr	co	fr	co	fr	co	
(i)	↑	↓	↓	↑	↓	↑	↑	↓	conventional
(ii)	↓	↑	↑	↓	↑	↓	↓	↑	sex-role rev.
	cu	pr	cu	pr	cu	pr	cu	pr	
	fr b	fr b	fr b	fr b	fr b	fr b	fr b	fr b	
(iii)	↑↑	↓↓	↑↑	↓↓	↓↓	↑↑	↓↓	↑↑	many, small
(iv)	↓↓	↑↑	↓↓	↑↑	↑↑	↓↓	↑↑	↓↓	few, large

Abbreviations and symbols: fr = frequency, co = costs, upward pointing arrow indicates high, downward pointing indicates low, cu = consumption, pr = production, fr = frequency, b = benefits. See text for full explanation.

conveniently the reverse of natural ones, then of course we would have a biased sex allocation.)

A more principled way to derive an equal sex ratio that also includes both natural and sexually selected components is to use differences in benefits related to ecology rather than costs. We assume a stable population so that the world is largely composed of species with more numerous, smaller or fewer, larger members. We also continue for convenience to assume that males have "quantity" consumption/production strategies while females have "quality" ones. In a population with more numerous smaller individuals (density low in size but high in numbers), then the male quantity consumption strategy would be more beneficial than the female quality one, but the female quality production strategy would be more beneficial than the male quantity one. Being at a higher naturally selected frequency, males are forced to compete for mates but eggs are forced to compete for sperm packets (Table 2, line (iii)). Alternatively, in a population with fewer larger individuals (density low in numbers but high in size), then the female quality consumption strategy would be more beneficial than the male quantity one, but the male quantity production strategy would be more beneficial than the female quality one. Being at a higher naturally selected frequency, females are forced to compete for mates but sperm packets are forced to compete for eggs (Table 2, line (iv)). In either case, assuming costs are equal, the population would be driven towards an equilibrium at equal male/female and egg/sperm packets ratios. Obviously, what is going on in any particular species is an empirical question. But the point is that there are good theoretical reasons for expecting the kind of diversity in the nature of gender differences and relations among species which have been uncovered in recent years (for some reviews see Kokko, Jennions and Brooks 2006, Clutton-Brock 2009; for a fascinating collection of cases see Judson 2002; and for a more controversial perspective see Roughgarden 2004).

More significant than the details is the meaning. On this view, gender relations biologically are literally a market, one in which an exchange of different naturally selected ecologically oriented functions take place in the first instance to the participants' mutual benefit, as Noë and Hammerstein (1994, 1995) suggested. While relations between isogametic mating types may conceivably, but not necessarily, be mutually exploitative or mutually cooperative based on economies of scale, and while there do exist a few species in which miniature males live exclusively as parasites of females, generally males should not be viewed as reproductive parasites nor should females be viewed as reproductive predators for that matter. An emphasis on the two-fold cost of sex (variously historically described as the cost of meiosis, of producing sons, or of male sexual competition to females), ignores the potential advantages of ecological specialization (but in general see Maynard Smith and Szathmary 1995, Corning 2005). As economists Coase (1937) and Williamson (1995) also made clear, however, markets have *transaction costs which is in part what sexual competition is* (although it would be helpful if the economics of transaction costs and the economics of conflict were more integrated with each other). When organisms turn from competing ecologically to competing sexually, they turn from optimizing in the narrow sense, i.e. ecologically, to playing social games. In a perfectly competitive market, we should expect these transaction costs or costs of antagonistic game playing to erode profits from specialization to zero. And that is exactly what we normally see in meiotic sex. The fact that only half of one's genes are placed in a sexual gamete as opposed to all of them in an asexual spore (there is a two-fold cost of sex) but that cost is normally exactly compensated for presumably by the advantages of ecological specialization so that meiosis is a two-division process from which four rather than two descendants emerge (i.e. there is equally a two-fold benefit). In some cases profits from the specialization and exchange above and beyond transaction and conflict costs remain – meiosis is followed immediately by one or more post-meiotic mitoses. In rare cases profits appear to be eroded below that – some eukaryotic protists have only a single-division meiosis – which not surprisingly is rare and it is unclear whether it is derived (as spite) or primitive. In other cases as in female animals, while meiosis is a two-division process, the divisions are unequal and two small "polar bodies" are abandoned – presumably a secondary derivation as an adaptation to the survival of rather than the numbers of grand-offspring.

I emphasize how varied the possibilities are for the naturally selected alternatives and hence for the forms that sexual selection may take concretely. For one thing, there are basically three kinds of markets in the human world – barter markets in which material goods and services are exchanged; currency markets in market economies in which material goods and services are exchanged for something of purely symbolic value, i.e. money; and financial markets in which things of symbolic value are exchanged for each other – money, stocks, bonds, etc. While I have emphasized here the material exchange which was probably at the heart of the origin of anisogamy as it was of markets in the human economic world, that does not mean that exchanges of the material for genes or even of genes for genes cannot have superceded this in particular cases.

Whether material, symbolic or both, sexual markets could pertain to consumption or production or both. Because the factors controlling population densities and frequencies measured in numbers and sizes may differ and even be negatively correlated, although males might eat more, through parental care they could make grand-offspring possible, and although females might digest more, they could be sperm limited. Sex role reversals in production are possible even at the cellular level. For example, although large, macrogametes could be responsible for making contact and thus for the production of offspring by secreting attracting pheromones. Although small, microgametes could be responsible for the production of grand-offspring through the provision of centrioles. Centrioles, the pair of barrel-shaped bodies that sit at right angles to each other at the poles of dividing cells, are inherited exclusively through males in most animals, while those from females are reabsorbed before or after fertilization. They replicate semi-conservatively through mitotic divisions but commonly separate to opposite poles for the second meiotic division. Hence if they were responsible for cell division, they would be responsible for the two-division meiosis. However, what role if any they actually play in cell division is unclear because the spindle controlling chromosomal division is actually nucleated from the "cloud of points" (periocentriolar material) which surrounds them. The differences may have nothing to do with density. In different species they could have to do with any of the other ecological conditions and strategy alternatives discussed in Chapter 3.

Even with those, the male provision of centrioles could be central to sex differences in animals. For example, centrioles also form the base of flagella. Hence females could be contributing growth while males contribute motility. Hence the use of dances, nuptial flights, etc., in intersexual selection. Through the aster, a brush-like structure that emanates from centrioles in many groups, centrioles are thought to organize the cytoskeleton, hence females could be contributing growth while males contribute maintenance. One important role for the centrioles is that they seem to determine the plane of cell division (cytokinesis – for discussions see O'Connell 2000, Lodish *et al.* 2000:834) Since development is a process of growth plus differentiation and developmental change, and since the plane of cell division resulting in offspring cells inheriting different cytoplasmic constituents is a major mechanism of differentiation in development, females could be contributing growth and males differentiation and developmental change (plasticity). Hence the use of symmetrical ornaments in inter-sexual selection (technically low fluctuating asymmetry). Females could be actively choosing for naturally selected high quality centrioles or males could be using sexually selected forms of them manipulatively to display symmetry in ornaments at the expense of that in other structures. One of, if not the most, thorough investigation of the function of centrioles available (in brown algae which are similar to animals in this respect) by Nagasato (2005) in addition to confirming much of the above including their role in cytokinesis concluded that ultimately the function of centrioles is related to sex differentiation.

Just as conflict can occur in a cooperative context, like children ganging up in a schoolyard fight or members of different human ethno-linguistic groups rallying to the side of their "brothers and sisters" in a conflict – cooperation can occur in an antagonistic context in gender relations. For example, in the context of sexual competition and selec-tion, cooperation can occur within a gender. Hence for example, groups of male kin or allies are known in a number of species to give every appearance of sometimes cooperating in coercing sex from females – in mallard ducks, bottlenose dolphins and right whales, for example. Similarly in leks, by displaying together it appears that groups of males cooperate in attracting females, and females in turn appear to cooperate by copying each others' choice of mate. Perhaps cooperation in an antagonistic context is less common than antagonism in a cooperative context because in the former case cooperation is based on doing the same thing while in the latter it is based on doing different things. The former may be less common than the latter because generally profits based on cooperation in the form of economies of scale are rarely as common or as great as those based on different specializations. That in turn may be the case because it is easier to cheat by just doing a little less of what everyone is doing.

In conclusion on gender relations, ideally a modern approach no longer assumes that gender relations are based on higher female than male parental investment. Rather, (i) it takes more, and preferably all major components of life histories into account, for example consumption as well as production, grand-offspring as well as offspring; (ii) it does (or should probably do more often than it does) relate the naturally selected components of different gender strategies to the ecological conditions under which they are favoured; (iii) it takes into account the benefits provided by these as well as the fact that costs may vary; (iv) it can or should use these to figure out the nature and degree of the social cooperation and/or conflict going on between genders; and (v) it integrates this understanding into the general framework of sex allocation theory (Charnov 1982). I am not aware of any empirical research that does all of these – but some (e.g. Fitze and Le Galliard 2008) come much closer than others.

4.6 COOPERATION

Just as there have been periods in western cultural history when humans have been viewed as uniquely bad (the emphasis on the "fall" in medieval Christianity) and periods when they have been viewed as uniquely good (the "heavenly city" of the eighteenth-century enlightenment; Becker 1932), there have been similar periods in the history of evolutionary thought. In the late 1960s, popularizations of ethology (defined then as the biology of behaviour) such as Robert Ardrey's *The Territorial Imperative* (1966), Konrad Lorenz's *On Aggression* (1967) and Desmond Morris' *The Naked Ape* (1967) emphasized the conflict-ridden character of human nature and hence of human societies. We are currently in the opposite kind of period when cooperation has taken centre stage in evolutionary theorizing (except in the study of gender relations!) and, not uncommonly, humans are viewed as a uniquely cooperative species. The past 25 years or so have seen a vast flowering of literature on the conditions favouring the evolution of cooperation in nature. Monographs and collections include Axelrod 1984, 1997,

2006; Casti and Karlqvist 1995; Maynard Smith and Szathmary 1995; Ridley 1996, 2001; Dugatkin 1997, 1999; Gadagkar 1997; Frank 1998; Sober and Wilson 1998; Keller 1999; Singer 2000; Wright 2001; Hammerstein 2003; Skyrms 2004; Corning 2005; Gintis *et al.* 2005; Wilson 2007; Blaffer Hardy 2009; and Okasha 2006.) The journal literature is even more vast – the Web of Science lists over 3000 articles on the evolution of cooperation, about half of them having been published in the past five years. For two reviews expressing different points of view see Nowak (2006) and West, Griffin and Gardner (2007).

This modern period of emphasis on cooperation began with the introduction by the late Bill Hamilton (1964a,b) of the principle that came to be known as "kin selection" or "inclusive fitness" mentioned briefly in the introductory chapter. Many believe that Hamilton's principle, along with his famous equation

$$br > c \qquad\qquad (4)$$

stating the condition required for the evolution of an altruistic act, one which benefits a recipient at some cost to the altruist, was the greatest innovation in the theory of evolution since Darwin. The meaning of the equation is fairly intuitive. Normally for a gene to be favoured by natural selection, its benefit in terms of producing mature offspring must be greater than its cost, hence $b > c$. This applies when the "donor" and "recipient" so to speak are the same individual. If the donor and recipient are different individuals however, Hamilton's equation inserts r, the coefficient of relatedness – r being the conditional probability that a gene responsible for the altruistic behaviour in the actor is also present in the recipient, identical by descent. Years earlier, J. B. S. Haldane had apparently declared that "I will jump into the river to save two brothers or eight cousins" (quoted in Nowak 2006 – actually, to be favoured by selection it would take a little more than two or eight). However, as Alfred North Whitehead is famously quoted, "everything of importance has been said before by somebody who did not discover it." Haldane apparently said it, but did not discover it at least in the way Hamilton did. As noted in the introduction, Hamilton's equation and initially suggested applications went on to give rise to a vast lineage of research on altruism among relatives in nature. It was this work which stimulated E. O. Wilson (1975) to write his *Sociobiology: The New Synthesis*.

While it is sometimes said that evolutionary models of cooperation including kin-selected altruism can be applied genetically or culturally (e.g. Nowak 2006), purely sociocultural applications of Hamilton's principle have been rare. However, in an empirical study of six years of editing and refereeing of the journal *Systematic Zoology*, Hull (1988: chapter 9) found that, consistent with the kin selection principle, these processes were at least somewhat biased in favour of those with related, and against those with unrelated philosophical, theoretical and methodological positions. Actually, more broadly the need for "identical by descent" has been questioned as Hamilton himself appreciated. Technically "green beard" genes in which the recognition of similarity only, illustrated hypothetically by Dawkins (1976) as recognition of those with green beards by those with green beards, not necessarily as a consequence of common descent, but still resulting in altruism, is theoretically possible and some cases are known. The important condition is the correlation between donor and recipient, however brought about. Social scientists too are not unfamiliar with the commonness not only of solidarity among kin, "nepotism", but of "homophily", the tendency to associate and cooperate with those like ourselves (for a review see McPherson, Smith-Lovin and Cook 2001).

In addition to kin-selected altruism and green beards, two other fundamental ways of thinking about the evolution of cooperation came to dominate thinking and research on the subject. These were reciprocal altruism (Trivers 1971) (usually understood as an exchange of the same thing at different times, sometimes divided into direct, indirect and network reciprocity, and on the larger scales particularly, commonly thought to require enforcement through punishment of non-cooperators) and group selection (roughly group-level positive effects outweighing individual-level negative effects). Stimulated in part by work in evolutionary biology on the subject, there has been a flowering of theoretical and experimental research in recent decades on cooperation in humans by social psychologists, behavioural economists and small group sociologists. Much of this research (summarized by Field 2001 and Fehr and Gintis 2007) has shown that, in a wide variety of situations, people behave altruistically much more than would be rational, in the sense of pursuing exclusively their own material self-interest although what process is believed responsible (biological evolution, cultural evolution, individual learning, rational choice?) is sometimes left vague. At the same time, it has also been shown to vary in degree culturally and is often thought (in all but small groups where repeated interactions are the norm and knowledge of reputations is readily acquired incidentally; Danchin *et al.* 2004) to require punishment (variously termed "altruistic punishment", "enforced cooperation" or "strong reciprocity") which itself is costly. Some of the most extensive recent modelling, however, suggests that costly punishment is only rarely likely to sustain indirect reciprocity (Ohtsuki, Iwasa and Nowak 2009 and by inference therefore to sustain network or "generalized" reciprocity as in Molm, Collett and Schaefer 2007).

The story of the controversies over group selection – conceptual, theoretical (including mathematical) and empirical – is too long and complex to recount here but has been well told by Okasha (2008). One of the most widely known theoretical results was Boyd and Richerson's (1985) that if cultural transmission is "conformist", i.e. positively frequency-dependent, so that individuals are likely to adopt the most common cultural variant, this decreases the variation within and increases the variation between groups thus increasing the strength of group selection and hence could be responsible for the evolution of cooperation among nonrelatives. While technically correct, the hitch is why positive-frequency dependence should obtain when negative frequency-dependence (avoiding competition) is more common. One answer could be cooperation based on economies of scale but then that assumes something, cooperation, which is what one is trying to explain with the derived group selection. Probably the empirical view most widely held currently is that of "multi-level selection" (e.g. Wilson and Wilson 2007). Life is functionally organized and hence shows evidence of selection on multiple levels of organization and these can include groups of individuals, as a minimum including eusocial colonies such as those of the ants, bees and wasps, for example. Here I would like to emphasize a possibility that has received far less attention than kin selection, green beards, reciprocal altruism or group selection – "byproduct mutualism", a term coined by Brown (1983) but which has come to be most associated with Dugatkin (1997, 1999).

Trees shelter birds, the birds' droppings fertilize the trees. As Lynn Margulis once remarked about symbiosis (intimate inter-species relationships) – organisms simply evolve to utilize each other's metabolic byproducts. Such "costless" cooperation in which the consequences of some act incidentally, as a byproduct of an otherwise "selfish" one, creates the conditions which favour another and vice versa is often thought to be theoretically uninteresting and even unlikely within a species. As Brown and Dugatkin have argued, however, neither is the case and I would like to emphasize the fundamental theoretical reason why that is so.

In a "scramble" for food or offspring production, what one individual manages to consume/produce others do not and vice versa. This is the case even if there is no shortage of resources or of places for individuals. Scramble competition is by definition a "minus–minus" relationship. The same obtains in "perfectly competitive" economic markets – what factors of production or of product sales that one firm gets, others do not and vice versa. However, keep in mind that at its base, genetic heredity in the form of base-pairs is a discrete phenomenon and that a number of fundamental ecological dimensions such as density, scale, patchiness, kind of uncertainty, etc., control the evolution of life histories including both consumption and production. The simplest possible alternative scenario to a scramble, then, is two alternative strategies in which the life history of one kind of alternative is adapted to some ecological dimension in one way, and the other in the opposite way. Where that is the case, the advantage is often said to be reducing competition, avoiding interference, etc. The same is said about "monopolistic competition" in economic markets – differentiating products and segmenting markets is advantageous because one obtains a monopoly, at least temporarily, in a particular segment. However, more than avoiding competition or obtaining a monopoly in a segment is required for such specialization to be advantageous. All of half a niche is no better than half of all a niche unless the specialization is accompanied by greater efficiency which it commonly is in the human economic and sociocultural world and so too, almost certainly, in the biological – i.e. there are positively non-additive fitness interactions among specialists. Under these conditions, *each alternative creates conditions which favour the other* – as has been emphasized here under density/frequency dependence – quantity consumption/production will favour quality consumption/production and vice versa. The advantages of specialization turn a minus–minus relationship into a plus–plus relationship, in short, into byproduct mutualism. Hence, on the basis of the simplest form of selectionist thinking, byproduct mutualisms should be almost the norm rather than a rarity within a population whether biological or sociocultural. If the alternatives then additionally engage in trade – then we have a form of reciprocity in which *different things* are exchanged, whether at the same or different times, rather than the conventional form of evolutionary reciprocity in which *the same things* are traded at different times. I have suggested that a material exchange of this type at the cellular level is at the heart of eukaryotic sex. This does not mean that typically "cheating" in the form of sexual competition and selection over the benefits of this exchange does not ensue – normally typically transforming a plus–plus relationship not back into a minus–minus one but into a zero–zero one.

At the same time, it is probably possible to view this interpretation of gender relations as readily from a group selection as from a byproduct mutualism perspective. The most recent and sophisticated analysis of the former is by Okasha (2006). A famous equation, the Price equation (1972), allows selection associated with character-fitness covariances on more than one level to be disaggregated, but such associations may not be causal. According to Okasha, however, with "contextual analysis" (essentially multiple regression), *cross-level byproducts* which *are* causal in nature can in theory be detected in both upward and downward directions between "particles" and "collectives". This would also seem to fit the case of genders and families as discussed above. Natural selection requires a positive character-fitness covariance. There is a positive character-fitness covariance in males and females which specialize ecologically (as opposed to those who do not). This is direct selection acting on individuals but which, when combined with recombination, as a byproduct, indirectly produces a positive character-fitness covariance in sexual families as opposed to the asexual. Similarly, there is a positive character-fitness covariance in families which recombine genes because of risk reduction (as opposed to those families who do not), but which, when combined with ecological specialization, as a byproduct, indirectly produces a positive character-fitness covariance among their members (at least overall or in the long run).

In human societies, the division of labour between females and males has varied and changed socioculturally historically enormously. For example in foraging societies it has often been between hunting and gathering and this has been proposed to have left its mark on our genomes in the form of the well-known greater average male ability mentally to rotate objects in space and the less well-known greater average female ability at memory for spatial locations as in the child's card game "concentration" (e.g. Tottenham *et al.* 2003). In pastoral and agricultural societies exchange was commonly organized on a larger scale in the form of marriage exchanges between groups and as classical anthropology and the great Claude Lévi-Strauss (1969) in particular showed, it has left its mark on semantic systems for naming and addressing relatives in languages around the world in which "parallel" relatives such as mother's sister and father's brother are distinguished from "cross" relatives such as mother's brother and father's sister. The subtleties of interpretation of such systems has been recently undergoing a revival (Allen *et al.* 2008; Chapais 2008). In the west for a couple of decades after the Second World War, the division of labour between genders was largely between consumption and production – males worked outside the home and women cared for children at home. A great revolution has taken place subsequently with the majority of women even with young children having moved into the labour force, the ramifications of which are still being felt. What appears to be happening in the west is that *any* cultural norms about this matter are disappearing and individuals and their families are working it out on their own – learning by trial and error in interaction with each other what works for them. However, theoretically, we should not be surprised that successful partnerships are likely based on some division of labour (commonly still some difference in commitment to labour force participation versus household tasks, as well as great diversity of arrangements within the latter.) On theoretical grounds we should also not be surprised, however, that these are only arrived at with some considerable conflict, albeit in a cooperative context, an interpretation which essentially combines conservative and radical views of gender relationships.

4.7 SUMMARY AND CONCLUSIONS

Rather than thinking of social interaction exclusively in terms of competition, which was historically the case in economics, or exclusively in terms of conflict and cooperation, which was historically the case in the other social sciences, biologists emphasize a tripartite distinction that suggests conditions under which they should obtain. With pure ecological competition, low densities/low frequencies favour "quantity" consumption/production while high densities/low frequencies favour "quality" consumption/production with a trade-off between resource depletion and environmental degradation. But low densities/low frequencies also favour pure ecological competition whereas high densities/low frequencies favour social interaction whether antagonistic or cooperative showing the intimate relationship between the latter. Similar distinctions apply to purely social strategies. Given those kinds of basics, however, concepts derived from the social sciences such as the advantages of specialization particularly under crowding, of portfolio diversification, of economies of scale and of the existence of transaction costs and those of conflict have contributed significantly to their further development and use.

To social scientists sex is biological and genders are sociocultural but to biologists sex is genetic recombination and genders are the biologically based differences and relations between sperm producers and egg producers. The great paleontologist and contributor to the modern synthetic theory of evolution, George Gaylord Simpson, once declared on the role of variation in biological taxonomy, "Variation is not incidental or an 'accident' to be ignored at any level in taxonomy; it belongs to the very nature of taxa and is part of the mechanism of their origin and continuing existence" (1961:50). The traditional evolutionary biological understanding of gender relations, i.e. sexual selection can be pretty well summed up by echoing Simpson: "Reproductive parasitism is not incidental or an 'accident' to be ignored at any level in gender relations, it belongs to the very nature of males and is part of the mechanism of their origin and continuing existence". This view of males as reproductive parasites is akin to a radical feminist view and cannot explain why females do not revert to parthenogenesis, to mating with other females, or to choosing males who do contribute resources causing the latter to evolve back to the allocation favoured by natural selection. One alternative is that in anisogamy, proto-males are naturally selected "quantity" consumers/producers while proto-females are naturally selected "quality" consumers/re-producers – with the former undergoing two mitotic divisions before sexual union thus providing the surface necessary to make contact and hence are responsible for the production of offspring while the latter fuel and direct the two meiotic divisions after sexual union and hence are responsible for the production of grand-offspring although many other possibilities including those pertaining to centrioles exist. In species with more extreme gender or kinds of gender-functioning differences as in familiar plants and animals, imbalances in costs/benefits of a variety of naturally selected functions add sexual competition and selection to the mix so that gender relations are literally a market, one in which the costs of transactions and conflict also enter in.

A variety of evolutionary theories of altruism and cooperation are available and cases of them known, including kin and green beard-directed altruism, reciprocal altruism in the form of the exchange of the same thing at different times and group selection. Experimental research by a variety of kinds of social scientists has shown that people, varying somewhat by culture, behave altruistically more than would be expected on narrowly rational grounds but not to an unlimited degree and why remains obscure. Although a favourite hypothesis, recent modelling suggests that costly punishment is unlikely to be the explanation. Fundamental principles of evolutionary ecology and the analysis of gender differences and relations suggest that the importance of byproduct mutualism as a consequence of specialization may have been underestimated. At the same time, the advantages of diversity under uncertainty on the level of families, populations and species suggests that higher levels of selection are also part of the story.

5

The ideal and the material: the role of memes in evolutionary social science

5.1 THE RECEPTION OF MEMETICS

The concept of evolution is central to any discussion of "memes". It was because of the possible existence of evolutionary processes beyond the gene-based biological that Richard Dawkins introduced the concept (suggestive of "m" for memory and "ene" of gene-like) as a possible substrate (1976:191–201, 322–31). Strangely enough he, of all people, did not initially clearly distinguish the gene and genome-like from the phene and phenome-like aspects of cultural evolution, a confusion which he corrected thereafter (e.g. 2003:119–127). The meme concept spread rapidly in popular culture. In his foreword to Blackmore (1999), Dawkins reported almost as many Google hits for "memetic" as for "sociobiology". However, it was generally not enthusiastically received in academic circles. Books on memetics (of which there have been at least seven – Brodie 1996; Lynch 1996; Blackmore 1999; Aunger 2000; Cullen 2000; Aunger 2002; Distin 2005) were interdisciplinary, which can itself be a problem. They often ignored many of the conventions of academic discourse, were sometimes written by non-professionals for a popular audience, and they were commonly viewed by social scientists, when they paid any attention at all, as yet another (post-sociobiology) incursion by biologists into their subject matter. In addition, a fratricidal war between adherents of the gene-like biologically adaptive view (which can most obviously be associated with vertical transmission) and adherents of the virus-like biologically maladaptive view (which can most obviously be associated with horizontal transmission) did not help when it should have been obvious that both are possible (see Chapter 9). The overall negative result was predictable – including among other things, the eventual demise of the *Journal of Memetics* (although it remains available on-line).

On the other hand, the reception among those interested in Darwinian-style theories of cultural evolution was more mixed. For example, Aunger was able to gather a group of academics including some well known ones including David Hull, Daniel Dennett and Robert Boyd for a conference that led to the anthology on "Darwinizing Culture" (Aunger 2000). Some of that interest has continued (e.g. see articles by Gil-White, Greenberg, and Chater in Hurley and Chater 2005). Moreover, a lot of those doing empirical and/or theoretical work on Darwinian-style cultural evolution in various social science disciplines often at least casually refer to memes. They commonly do so because it helps to distinguish what they are doing from sociobiology/ human behaviour ecology/evolutionary psychology as well as from the developmental stage theories of history traditionally called "evolutionary" in the social sciences. Examples that come to mind are Van Driem's (2001) symbiotic theory of language and some of the essays in Lipo et al. (2006) on archaeology and prehistory. And of course, the internet generation takes the concept of memes – roughly "ideas which spread" for granted (but see Section 5.5 below). As a colleague put it to me, "the concept is out there" (my local newspaper publishes a "meme of the week") and it gets, and undoubtedly will continue to get picked up and used in interesting and surprising ways. For example Keith Stanovich, an accomplished cognitive psychologist, argued in *The Robot's Rebellion: Finding Meaning in the Age of Darwin* (2004) that in pursuit of humanistic and democratic values, we (the robots of the title) need to bootstrap our way to rebellion against both our genes and our memes.

Given that in addition to Dawkins himself, some of the greatest evolutionary biologists of our time including George Williams (1992:15–16, 18–19), John Maynard Smith (Maynard Smith and Warren 1982; Maynard Smith and Szathmary 1995:309) and Paul Ehrlich (2000) have made clear their awareness of the significance of a Darwinian cultural evolutionary process, and some, including Luigi Cavalli-Sforza and Marcus Feldman (1981, and subsequently for Feldman) have even made it a major part of their work, one might have expected biologists to display more enthusiasm. There currently are some biologists working on cultural evolution (e.g. Mesoudi, Whiten and Laland 2006) albeit with little or no reference to memes. I believe there is a reason why such work is not more widespread among biologists.

The meme concept was introduced just at a time when there were rising "discontents" (Ruse 2006) within the biological community with neo-Darwinism (as it was known in Britain), or the synthetic theory of evolution (as it was known in America), i.e. with population genetics or the genetical theory of evolution. Those discontents included an implicitly naive view of the origin of life; an extreme micro and gradualist emphasis; an overemphasis on conflict as opposed to cooperation; a relative neglect of development and ecology; and overly restrictive theories of speciation and macroevolution. Moreover, it was introduced by the very person, Richard Dawkins, around whose work many of those discontents crystallized. Nevertheless, it is fair to suggest that, by its linkage in people's minds, the wide diffusion of the meme concept gave Darwinian-style cultural evolution a lift, helping move the latter some distance out of the small, scattered academic niches in which it dwelt at the time.

5.2   BIOLOGICAL GENES

Beyond the sociology of its reception, objections to the meme concept that there are discrete units of symbolically encoded biological information which evolve – genes, but not of cultural information – memes, are not persuasive. A fair amount of space will be devoted here to talking about genes. I think that may be useful because memes are intended to be analogous to genes, and the gene concept itself is commonly taken for granted in the memetics literature in a way that is problematic.

> Genetic units of structure, function, replication and recombination do not coincide with one another. Units of structure include base pairs, nucleosomes, 30-nm fibres, loops, and chromosomes. Units of function include codons, the traditional molecular 'gene' of a DNA sequence coding for a single strand of a protein molecule, as well as many, many potential others e.g. with introns counted in or out, adjacent and even distantly acting regulatory sequences in or out, sequences coding for other strands of the same protein in or out, sequences coding for other enzymes functioning in the same pathway in or out, and ultimately even whole hierarchies and networks serving some particular ecological, sexual or social function. Units of replication are replicons and chromosomes. Units of recombination can be sequences of a length 'short enough to be different and long enough to make a difference' as the popular version of Williams (1966) definition has it in crossing over, and are chromosomes in independent assortment... This lack of precise correspondence and consequent multiple 'gene' concepts has been the source of endless angst in the history of biology (Blute 2005:404).

The well-informed may notice that I did not include one of the most common gene concepts, ORFs (open reading frames, i.e. between initiator and terminator codons) in the functional list. That is because they are units of intermediate function (transcription into RNA) but not necessarily of ultimate function (translation into protein). The essential point, however, is that not only have concepts of what exactly a "gene" is changed historically (for a brief history see Rheinberger and Muller-Willie 2004; for more in-depth discussions see the essays in Beurton, Falk and Rheinberger 2000), they also vary currently among biologists as survey research among them has shown (Stotz, Griffiths, and Knight 2004). Some biologists even blog about it (Moran 2007). This variation and change is of course exactly what a cultural evolutionist would expect, including in science (Hull 1988).

One can sometimes get the impression from recent literature that this was a new problem with the coming of the molecular biological discovery of "genes in pieces" and all that but that is not the case. As early as 1957, Seymour Benzer was advocating, on the basis of results from his elegant experiments on mapping within classical Mendelian genes and interpreting them in terms of the new DNA theory and later of the Watson and Crick model, that the term "gene" be replaced by cistron, muton, replicon, and recon for units of function, mutation, replication and recombination respectively (Holmes 2000). And from then, at least through the 1960s and early 1970s, the problem was a topic of discussion even in textbooks (for example in the several editions of H. L. K. Whitehouse's *Towards an Understanding of the Mechanism of Heredity* where I first read about them in the early 1970s).

Although it was not discussed in *Adaptation and Natural Selection*, this was the background against which George Williams classic book was written. Which, if any, of Benzer's genes is it, that evolutionists in general, and population geneticists in particular, are talking about? It was apparent to Williams that it was none of them, that evolutionists needed their own "gene" and hence his definitions: "that which segregates and recombines with appreciable frequency" (1966:24) and "in evolutionary theory" "any hereditary information for which there is a favorable or unfavorable selection bias equal to several or many times its rate of endogenous change" (1966:25). Because the point of Williams' definition was that the evolutionary gene needed to be short enough to tend to remain intact through recombination as well as be long enough to affect function and hence be subject to selection, his gene came to be known as that which is "short enough to be different, and long enough to make a difference" (origin unknown to me). These are impossible criteria to combine. If being a gene is both a negative and a positive function of sequence length, which on the simplest assumption combine additively for example, then no sequence – short, intermediate or long – would be any more gene-like than any other. Later, in distinguishing between the "domains" of information and matter ("codical" and "material" domains respectively), emphasizing that a gene is "a package of information, not an object", he reiterated the theme that to evolve by natural selection, a "given package of information (codex) must proliferate faster than it changes" (1992:11). He also noted that the same thing is true of memes (1992:13).

In 1976, between these two books of Williams, Dawkins introduced the term "replicator" in his classic, *The Selfish Gene*. Not only did it carry the unfortunate connotation that genes replicate themselves instead of being replicated by enzymes, imply that the origin of life problem was pretty much one of "once upon a time there came a replicator", and lent itself to misunderstanding by uninformed readers that the primordial replicator might possibly have been DNA, but it also tended to suggest the structural (and hence the short) rather than the functional (and hence the long) component of Williams' definition. Later, in *The Extended Phenotype*, subtitled *The Gene as the Unit of Selection* (1982), Dawkins got more specific by specifying the criteria of copying fidelity, longevity, and fecundity. Note how the copying fidelity equates with the "short enough" component of Williams' definition and the longevity and fecundity (i.e. viability and reproductive success, the two most inclusive components of fitness) equate with the "long enough" component of Williams' definition. (To further confuse the waters, however, he also called it the "optimon" which suggests the latter, i.e. that which functions, whereas replicator tended to suggest the former, i.e. that which maintains its structure!) It probably would have been better if both Williams and Dawkins had gone to "evcon" which, like Mayr's "selectron", rhymes with cistron, etc., but, unlike it, does not unduly emphasize function over maintaining structure (see discussion in Dawkins 1982:81).

Whatever term is chosen and whatever description of the criteria is used, the important point is that the evolutionary gene combines a unit of recombination (which is the unit of genetic transmission in sexual species) with a unit of function. However, since these do not in fact always or perhaps even often coincide – the *evolutionary* gene or replicator is as much a fiction, a theoretical construct rather than a hypothetical entity, as is an ideal gas for example. Williams understood this. He often spoke of "hereditary information" instead of genes, sometimes put "the gene" in quotations marks, called it the "abstract" gene of population genetics (1966:24), and noted that such a gene "would produce or maintain adaptation as a matter of definition" (1966:25). Such fictions are useful one must hasten to add. One can write equations and prove theorems about them just like one can about right angled triangles say, which is what is done in theoretical population genetics. To the extent that things in the real world approximate to the axioms employed (and they *always* only approximate), then what can be proven deductively true about the ideal entity will also tend to be true about the objects in the material as opposed to the conceptual world. In the real, as opposed to the fictional, world, because recombination commonly does not follow functional lines, the relationship between units of transmission and function is not always one-to-one but can be one-to-many or many-to-one. This is the case not only in the traditional sense that one Mendelian gene can affect many traits and many Mendelian genes can affect the same trait, but also on a finer scale so that information from the same smaller genomic sequence can end up incorporated into more than one polypeptide chain and one polypeptide chain can be woven together from the information in scattered genomic sequences. Such cutting and pasting can and does take place in a variety of ways at any stage from transcription to post-translation.

Modern evolutionary theory includes a variety of responses to this dilemma. Theoretical population genetics does so by subtly altering the meaning of the term "gene" from elementary expositions in the early chapters of textbooks implying a unit of transmission and function both to more advanced expositions in later chapters implying a unit of transmission only. There the "many to one" and "one to many" problem is handled by the statistical theory of quantitative genetics developed originally in agricultural applications. If the (however) many genes influencing a trait combine additively to influence that trait or if the (however) many traits influenced by a gene combine additively to affect fitness, then such effects will contribute to the correlation between parents and offspring, i.e. will be heritable and, leaving aside sampling error, will be expected to respond to natural selection. Note, however, it is only the *additive* genetic variance of fitness that is heritable and therefore available to natural selection – a theoretical result established long ago by Ronald Fisher. Only then should adaptation on the level of observable characteristics of organisms be expected; only then should *material* organisms, as opposed to *ideal* information-containing genes, be expected to optimize and play games with each other rationally because of the influence of natural selection. However, we simply do not know if the additive assumption is correct. For example, Hill, Goddard and Visscher (2008) have recently presented both empirical and theoretical arguments why it normally does obtain. O'Hara (2008) on the other hand has politely questioned the theoretical analysis at least.

At the opposite extreme, evolutionary (including behavioural) ecology which utilizes optimization and game theory models in the study of development, physiology and behaviour by experimental, observational and comparative methods has restricted their units to function rather than transmission – albeit abandoning the term "gene" for the term "strategy". Despite their critics, who are legion, as we saw in Chapters 3 and 4, evolutionary ecology is one of the two most active and productive research programmes in evolutionary biology – the one yielding most of those fascinating insights into animal behaviour that one reads about in one's newspaper almost daily. Practitioners of this "adaptationist programme" have voted with their feet in part because, as a practical matter, the complex genetic basis of most characteristics, particularly most behavioural ones, in most species are not, and never will be known. At the same time, this "phenotypic gambit" (Grafen 1991) as it has been called has its own risk which corresponds to the population geneticists' risky additive assumption. If because of recombination, the phenotypic effects of a particular set of interacting genes may not be heritable, by the same token, a particular material phenotype may not have a simple enough genetic basis to be heritable. Without the additive assumption, neither starting point taken alone yields the correlation between parent and offspring required for evolution by natural selection.

In between these two ideal and materialist extremes, the dilemma is sometimes temporarily papered over verbally. For example, Hull (e.g. 2001a: chapter 1) has argued that the debate over "the units of selection" is misguided because the predicates involved are ascribable to different entities. According to Hull, evolution is about replicators (which vary), interactors (which are selected) and lineages (which evolve). These may be, but are not necessarily, genes, organisms and species respectively. Terminologically, there seems to be little choice other than to tolerate the variety of gene concepts used in particular contexts in particular research traditions in biology, which is what many philosophers and biologists who have considered the question have concluded and what the majority of biologists in fact do (Knight 2007). As an example of how the saga continues, the website of the HapMap project calls SNPs "alleles" (i.e. the historic term for different versions of a gene). So a single nucleotide is now a gene! A second alternative would be to abandon the term gene completely and use the various historically suggested "on" – recon, cistron, etc., substitutes. Not only would concepts of function need to be multiplied further given current molecular knowledge, but the approach seems sociologically quite unlikely to be successful, given that the transition did not take place in their heyday. As all evolutionists know, there is an inertia to history. A third possibility is to do what Burt and Trivers (2006) largely do in their monograph on intra-individual conflict, which is to talk about "genetic elements" or sequences, or even better, genomic elements or sequences, all of which conservatively retain the historic "g" word or its root while signalling more cognizance of the complexities involved. Taking cognizance of the existence of inheritance beyond the gene, one might want to take the even more radical step and speak only of "hereditary elements".

The really interesting question is not just how to cope verbally with the fact that the kinds of units required for relatively accurate transmission on the one hand and function on the other hand do not always coincide, but *why they don't*. Why has recombination which breaks up favourable "gene combinations" (as it is said) in every generation evolved and been maintained? That, after the origin of life, remains the mystery of mysteries. Whatever the merits may or may not be of the various possibilities, the point is that the evolutionary version at least of the "lost gene concept" from the title of Griffiths' (2002) review of Beurton, Falk and Rheinberger (2000) is lost in the territory of recombination and his "reward to finder" will go to those who search there.

In addition there is a second confounding issue. The material as well as the ideal can be inherited. DNA is only a small proportion by dry weight of what a cell or organism inherits. Much inheritance, particularly through females, is cytoplasmic. In addition to DNA, a cell or organism may inherit RNA (managerial knowledge and expertise), mitochondria (power plants), ribosomes (assembly lines), membranes (buildings), enzymes (tools) and a whole host of other "resources" (Oyama 2000; Jablonka and Lamb 2006).

Having disabused the reader, if you were not already, of the notion that the same "gene" is both a unit of recombination, i.e. of transmission in sexual species, and a unit of function – one perhaps coincident with the molecular biological concept of that which codes for a single polypeptide strand and that only genetic information is inherited – where does all of this leave memes? In order to answer that question, another, also somewhat lengthy, preliminary discussion is necessary, on the role of language in cultural evolution.

5.3   LINGUISTIC MEMES

Most cultural transmission in the human species takes place using language in whatever medium it is embodied – gesture, sound, print, electronic, etc. Presumably it is because we can *tell* our children (and others) what we know and what to do, rather than just *show* them, that partially accounts for the great success of our species. Ten points about genetic and linguistic systems of transmission are basic here.

(i) Both genetic and linguistic forms of heredity are digital at their base (nucleotide bases and phonemes respectively) which facilitates stability in transmission (Dawkins 1995). So the problem is not that genes are discrete and culture, at least in linguistic form, is not.

(ii) In both genomes and language there are basic units of function as well as of structure (codons and morphemes respectively) and in both there are a number of more inclusive units of each.

(iii) In both, units of function are said to be "arbitrary", "conventional" or "symbolic" because of the nature of the link between symbol and what it stands for or represents (codons for amino acids biologically; morphemes and lexical items for their reference linguistically). In origin, the genetic code was probably more iconic than symbolic (as was the case in the early history of languages as evidenced by the history of written languages, discussed for example in Burling 2005: chapter 6). Currently, however, genetically encoded information is legitimately said to be arbitrary or conventional or symbolic. There is no physio-chemically necessary connection between a genetic codon (a triplet of nucleotide bases in DNA) and what it stands for or represents (an amino acid in a protein molecule). According to current knowledge the connection is as much a product of evolutionary history as is the connection between red and stop rather than go, or between the word father and a male parent rather than a female one. It could have been the opposite (Crick 1968). (This is possible biochemically because "adaptor" molecules are what associate a genetic codon, translated into messenger RNA on one side, and the amino acid component of a protein it stands for, represents, or encodes for on the other side. These adaptor molecules – tRNAs recognize the message with their anti-codon, and aminoacyl-tRNA synthetases recognize both the tRNA and the protein component and link them – have themselves evolved –

they are a product of evolutionary history.) To be sure the meaning of "no necessary connection" in the two cases is substantively different. Biologists say there is "no necessary connection" meaning no physio-chemical necessity (Stegmann 2004, after Monad). Linguists say there is "no necessary connection" meaning in addition that there is no biological or psychological necessity. Nevertheless the point is theoretically ultimately the same. The association between symbol and what it symbolizes is a fact of some evolutionary history rather than a necessity.

"Semantic talk" in biology has been subject to criticism in the last decade (for an overview and references see Godfrey-Smith 2007), much of which I believe to be generally misplaced. For example, even Stegmann (2004), who clarified for philosophers that ultimately the "arbitrariness" of the genetic code is attributable to the fact that the particular combination of binding sites on a macromolecule (e.g. on a tRNA) is not a physio-chemical necessity, maintains that fact does not justify the attribution of semantic information to nucleic acids which is correct up to a point. The attribution of semantic concepts requires that in addition, we turn our attention from where the association does *not* come from to where it *does* come from – a history of a selection process.

Consider an analogous psychological example. If we teach a rat with food as a reinforcer to turn left in a T-arm maze when a light is on and to turn right when it is off, we call the light a signal that has acquired *meaning*; the rat has learned to *expect* food to the left with the light on, and food to the right when it is off. There is no biological law nor even a universal law of rat psychology that a light means if you turn left you will acquire food and no light means if you turn right you will acquire food (and assuming the rat is hungry that it will behave accordingly). Rather the light has acquired that meaning through a history of reinforcement-type selection. Similarly, despite the fact that the *current* functioning of the components of cells like nucleic acids and proteins could in theory be given a complete description in terms of the chemistry of molecular recognition, enzymatic action and so on, such a reductionist description would be utterly incapable of providing a broader appreciation of the historical evolutionary context tha t is the reason for their existence. In addition to principles of physics and chemistry, understanding why requires additional biological principles such as those of natural selection, drift, etc.

Nuclei may best be viewed as the brains of cells which perceive, calculate and act (Blute 2005). The information in their genomes represent the cell's encoded memory of events in the past history of its lineage (including mutation, drift, selection, etc.) and, by the same token, their expectations of and hence basis for action in the future. More broadly, cells and organisms have genomes presumably for the same reason vertebrates have brains. Replete as they are with "if statements" governing routines, sub-routines and so on, genetic programmes make possible a complexity and flexibility in development, physiology and behaviour vastly greater than would be possible without them. None of which is to deny that much more than genomes are inherited, i.e. that much development, particularly early, is maternally programmed.

Moreover, it is sociologically impossible to imagine a wholesale revision of the language of information, the code, synonymous codons, transcription, translation, proof reading, editing and so on in molecular biology taking place anytime soon. Indeed, "semantic talk" has been expanding rather than contracting (Searls 2002). A couple of more recent introductions in genomics are "annotating" sequences (a way of lumping all considerations of function together) and the search for sequence "motifs", variations on which are characteristic of some particular phenomenon.

(iv)   In neither the genetic nor the linguistic case do units of structure and function necessarily coincide; in particular the smallest unit of structure (nucleotide base, phoneme) is smaller than the smallest unit of function (codon, morpheme).

(v)   In both, the building of larger units of function from permutations of smaller units of structure (e.g. of codons from nucleotide bases, morphemes from phonemes) is what makes possible "unlimited inheritance" (Maynard Smith and Szathmary 1995) and thereby facilitates adaptation by means of the cumulative evolution of complex, diverse and therefore unique, as well as flexible entities in both cases. In both, however, much remains to be learned about the relationship between structure and function because in both many units serve purely internal organizing and controlling functions (*cis* and *trans* acting regulatory sequences as well as probably a lot of the untranslated RNA in cells; many morphemes and words in language which serve purely grammatical functions). Note that the term "functional" is used in the opposite sense in the two disciplines – e.g. for the purely grammatical in linguistics.)

(vi) Eventually in both one arrives at that which is capable of "standing alone" up to a point (a genome, and traditionally an utterance or a sentence). Then of course there are populations or species of such with variation among individuals. We all say something a little, and sometimes a lot differently, even in similar situations. (If we were talking about the evolution of *languages* rather than other aspects of culture *expressed* in language such as hip hop, scientific or Serbian culture for example – then these populations and species would be of the idiolects of individuals. A language is a species in which members are able to exchange communications linguistically within its boundaries but not beyond them analogous to a biological species with members able to exchange genes within but not beyond its boundaries.)

(vii) The famed recursiveness of language (sentences within sentences, clauses within clauses, etc., making possible virtually unlimited expansion) may, as has been claimed by Hauser, Chomsky and Fitch (2002), or may not, in the long run, turn out to be unique to human language among animal communication systems. It does not, however, appear to be universal among the former (Everett 2005). It is probably also not unique in another sense in that it represents the major means by which something new is generated in evolutionary processes including the genetic – by duplication, insertion and divergence on multiple levels. The understanding of this in genetics goes back to Ohno (1970).

(viii) Grammaticization (in the narrow sense by which sense and meaning are emphasized or expanded by additions but which then tend to become more economically expressed by the use of purely grammatical elements both morphologically and syntactically) is a process apparent in both realms. It is most obvious genetically in the great increase in the proportion of regulatory DNA in the transition from prokaryotes to eukaryotes. What remains unclear in language is how much grammaticization took place in the biological evolution of the human capacity for language (thus confirming the structural linguists' dream of a universal grammar); how much in the cultural evolution of languages plural (thus confirming the historical and anthropological linguists' view of the cultural evolution of grammar); or even how much takes place in the acquisition of language by individuals (thus confirming the developmental and social psychologists' view of grammatical acquisition in development).

(ix)   When culture is transmitted linguistically, there is no systematic recombination process every generation – which is not to say that recombination does not happen. Much of the famous "productivity" of language occurs as a result of such "reslotting". We can say the boy ran "up the hill" or "across the road" or "over to her", etc. However, as with genomes, recombination in language can and sometimes does take place without regard to functional units, which often destroys the result semantically including the "sense" of a component and the "meaning" of a whole. These terms roughly correspond to "function" and adaptive status or "fitness" in biology respectively and the entities of which they are properties roughly correspond to the terms "meme" and "memeplex" in memetics (see the discussion in Blackmore 1999: 18ff). If a meme does not make "sense" (have a useful function) or a memplex is not "meaningful" (not adaptive in the setting in which it is found), they are unlikely to be transmitted further. Memes do not always work well together in memeplexes – we have all struggled to make "sense" of what someone else "means". The potentially pathological effect of recombination is seen at its most extreme in the "word-salad" of classic schizophrenic language in which syntax can remain eerily normal while meaning is destroyed.

"If we need soap when you can jump into a pool of water, and then when you go to buy your gasoline, my folks always thought they should get pop, but the best thing is to get motor oil" (quoted in Covington et al. 2005).

Similar phenomena can take place in units smaller than sentences so that schizophrenics commonly coin neologisms which are morphologically correct in that they could be a word in the language – except that they are not, as well as in larger units – so that an entire narrative can be "florid" as it is described, essentially not conveying intelligible meaning. On the other hand, as in genomes, recombination can also be creative. Sometimes when we struggle to understand what someone else means, that may be because they are saying something new, something which may well turn out to be worth hearing. Some neologisms, "meme" itself is an example, *are* picked up and widely disseminated.

(x)   In conclusion to this section on the role of language in cultural evolution, what it all says to me is that to the extent that human culture exists in linguistic form which it largely does, *the "meme" concept is no more (but admittedly no less) problematic than is the gene concept.* Hence it cannot be banished on a priori grounds.

5.4 THE SCIENTIFIC USEFULNESS OF THE "MEME" CONCEPT

Not surprisingly, as with the gene concept historically, despite the complexity, in some cases "memes" can be shown to have done a useful job of scientific work. Here we will consider three examples – Pocklington and Best (1997) on internet posts, Ritt (2004) on phonological changes in English, and Jan (2007) on the string quartets of Haydn, Mozart and Beethoven. Inspired by the meme concept, Pocklington and Best (1997) used a text retrieval algorithm (latent semantic indexing) to identify sets of rare words that co-occur (sets which they called "term-subspace traits") in posts to some news groups on the internet. Since many posts originate in response to previous posts, they are threaded, i.e. the authors possessed genealogical data. They were able to show statistically that in some cases the reproductive success of a post, i.e. its success in generating in-reply-to posts, was a function of the degree to which certain of these traits were expressed within it. There was some lack of clarity in terminology (they sometimes referred to a term-subspace as a replicator, other times as a trait, and still other times as an indicator only of the underlying "cognitive motif" which is the true "conceptual replicator" although later they seemed to settle on the set of words that co-occur – Best and Pocklington 1999). Despite that, the study clearly was a proof-of-principle illustration of cultural microevolution in texts.

But also note two things. A set of words that co-occur is not strictly compatible with the evolutionary gene or replicator concept because the words are not necessarily adjacent to each other or "linked" in genetical terms. For example in one case illustrated the words "James Smith" and "Nazi" were not necessarily found side-by-side. The relationship between structure and adaptive function, then, was many-to-one. (If you are wondering how that is possible, an evolutionist would explain that such co-occurrence without linkage is possible because, in addition to their ecological effects, separated sequences can have social effects on each other, i.e. have positively or negatively non-additive effects on fitness due to their co-occurrence. Moreover, with positively non-additive fitness effects, an evolutionary geneticist would tend to expect them to evolve towards becoming linked making it less likely that they become separated.) Secondly, in some cases the adaptive effect of the terms on the post could have been the result of someone saying "James Smith is a Nazi" but in other cases it could have been the result of someone saying "How dare you call James Smith a Nazi" (an issue which the authors subsequently addressed, Best and Pocklington 1999). The relationship, then, was many-to-one. In additional work, by grouping posts similar in their term-subspace contents into quasi-species, Best (1997) was also able to show that similar posts tend to be related by descent (come from the same thread) and that competition among quasi-species is more intense the closer their ecological niche.

A second example is provided by the work of Ritt (2004) on phonological changes in English. Old English was apparently under conflicting constraints which therefore prompted changes in Early Middle English. One constraint was "fixed word stress" (i.e. the syllable stressed in a word, with rare exceptions, was constant irrespective of context). A second was at the same time, the temporal interval between stresses in an utterance was preferred to be constant. The most obvious solution to this conflict is compression and lengthening of syllables or words. Notice the compression in Present Day English of "syllable" and the lengthening of "foot" (roughly a group of syllables which can be a word) in the following utterances which serve to maintain a constant temporal interval between stresses.

> Every / syllable / has a / rhythm
> Every / foot   / has a / rhythm

Because vowels are inherently more flexible than consonants, compression and lengthening is typically of vowels.

> He is a / good   / friend of / mine
> He is a / good old / friend of / mine

Hence Early Middle English was characterized by a series of "quantity adjustments" (vowel lengthenings and shortenings) as well as of "quality adjustments" (replacements among long and short vowels) (chapter 8). These adjustments were part of a longer term trend in the evolution from Germanic to Late Middle English in which words converged on a trochaic-like structure of feet – what Ritt calls "the great trochaic conspiracy" (chapter 9). (A trochee is a metrical foot in poetry composed of two syllables, a stressed one followed by an unstressed one. Longfellow's poem, *The Song of Hiawatha*, of which the first line is, "**Should** you **ask** me **whence** these **sto**ries", is written almost entirely in trochees.) Rather than categorical rules, Ritt argues that all of these changes can best be explained as examples of the adaptive evolution of memes coding for vowel phonemes and for the morphotactic shapes of words to memes coding for the regularities of English utterance rhythm. Linguistic memes (which include a variety of features or constituents of linguistic competence), just like genomic elements, evolve not only in adaptation to the ecological environment, but also *in adaptation to each other*. Note however that Ritt's examples of the variety of linguistic competencies (chapter 6) include a mix of structural (e.g. phonemes) and functional (e.g. morphemes) elements. This does not, however, negate his larger point that memetics, as a token of the type "Darwinian evolutionary theories", suggests a non-teleological explanation of the functionality of the kinds of changes in language described.

In a third case in a series of articles (e.g. 2003) and a book (2007), Jan has argued that a number of hierarchical levels of stylistic memes are identifiable in the structure and evolution of music. These begin at the bottom with configurations of pitch and rhythm and run all the way up to the large-scale archetypes of western music. These are identifiable, whether in the same piece, in different pieces by the same composer, or within the output of an historically or geographically distinct group of composers. He introduces some notation for the representation of the memetics of music (2007:chapter 3) and advocates quantitative methods in their study – suggesting the use of David Huron's *Humdrum Toolkit* software for this purpose (2007:chapter 6). He illustrates the use of the latter in this context with a case study of descending chromatic tetrachords of which he uncovers a number of "alleles" (allomemes) in string quartets of Haydn, Mozart and Beethoven. The analysis is about similarity rather than shared, derived similarities and much of the same tension over "particulatness" and the structure–function dichotomy is evident in this work as elsewhere in memetics. Jan tends to place the emphasis on structure. For example he states: "Music . . . has the propensity of having no fixed semantic structure, but a relatively stable syntactic structure" (2007:26) and scolds Dennett for claiming that three notes cannot be a meme because it does not make a melody (2007:55). Despite such issues, Jan's memetic approach obviously moves the study of "intertextuality" in music a significant step forward and his larger critique of explanations based on the conscious agency of composers and suggested replacement with ones based on "vast meme-complexes running like software on the physical platform of the human brain" (2007:xiii) is at the very minimum, a provocative innovation in musicology.

5.5   ALTERNATIVE POPULAR USAGE AND TERMS
IN EVOLUTIONARY SOCIAL SCIENCES

I used Pocklington and Best, Ritt, and Jan's work to make the point that despite the sometimes same structure or transmission versus function disjunction problem as in genomes, the concept of a meme, like that of a gene, can and has historically been used to do useful scientific work. It is worth noting at this point that academic and popular usage of the "meme" concept can be at variance with one another. Rather than a unit of information that spreads, one popular meaning is closer to what could be termed an electronic viral "phemome" (analogous to a biological phenome, see terminology in Table 3). According to my unscientific survey of some undergraduate students in my geographic area – the term meme is well entrenched in youth culture but first, is restricted to communications received electronically. So a joke being passed around on the Internet is a meme, but one being passed around by word of mouth is not. Secondly, to them a meme is not analogous to a gene but to an organism – not a *unit* of information, but the *manifestation* of an entire functionally integrated *package* of information – i.e. the text, photo, song, video, etc., that they see on their computer screen and/or hear from its speaker. Thirdly, their meme is viral-like, not in the sense of normally or even usually being biologically harmful, but in the sense that the bits that arrive electronically to their

Table 3. *Terminology for units of genetic and memetic information and their manifestations*

	Biological	Sociocultural
Information	Gene (genetic)	Meme (memetic)
	Allele (allelic)	Allomeme
	Genome (genomic)	Memome (memomic)
Manifestation	Phene (phenetic)	Pheme (phemetic)
	Phenome (phenomic)	Phemome (phemomic)

In this table, nouns are entered first, adjectives are in brackets. At least seven of the biological terms in column one are in wide usage – gene, genetic, allele, allelic, genome, genomic, and phenetic. Three less widely used are filled in – phene, phenome and phenomic. Three of the sociocultural terms in column two are in wide usage – meme, memetic and allomeme. Suggested alternatives for the rest are filled in.

computers have to be translated into the text, video, etc., of what they mean by a meme. As with a virus and its host, the apparatus required for that translation does not come with the transmission, but is instead part of the receiver's computer.

So while the concept of a meme, like that of a gene, can and has historically been used to do useful scientific work (and is well entrenched, albeit with a somewhat different meaning, in youth popular culture), it is not the only word to have done so. A lot of terms other than memes have been historically, and are currently, in use in different social scientific disciplines for the "iss and oughts" normally expressed linguistically and which evolve culturally. Mesoudi *et al.* (2004, 2006), paraphrasing Boyd and Richerson (1985) and Richerson and Boyd (2005) use information – "information such as knowledge, beliefs, and values that is inherited through social learning and expressed in behavior and artifacts" or "information capable of affecting individuals' behaviour that they acquire from other members of their species through teaching, imitation, and other forms of social transmission." "Information" is usefully ambiguous – implying something that can be transmitted *and/or* something that can affect behaviour. Sociologists tend to refer to "norms and values", linguists to rules or "competencies", institutional including evolutionary economists and organization theorists to "conventions", "habits", "routines" and "competencies". Anthropologists once liked to talk about "folkways and mores" but today more often say "traditions" (albeit some students of animal behaviour like to call animal cultures "traditions" to distinguish them from human culture). Archaeological speak is quite varied – "techniques", "design elements", "traits" and "traditions" for example are fairly common. In science studies they speak of concepts, theories and methods although sometimes more inclusive entities such as "research programmes" and "paradigms" are investigated.

It is not necessary, nor is it likely, that those doing cultural evolutionary research in such a wide variety of social science disciplines should or will abandon the conventional terminology used descriptively for their particular sphere of culture by evolutionists and non-evolutionists alike in favour of memes, memeplexes and the like. Whatever terminology is used, it is important that knowledge not only in the sense of what to expect but also in the sense of what to do be included. Both biology and culture include not only knowledge about the world but also how to act in it. As noted previously, however, talk of memes has commonly been found useful for communicative purposes in this wild west of sociocultural evolution when the discourse is interdisciplinary.

5.6   MEMES AND SOCIAL LEARNING
MECHANISMS

Most social communication in animals, and perhaps in people as well, is not about social learning in the sense relevant to culture, i.e. about Boyd and Richerson's (1985) "second inheritance system" that creates similarity between individuals. More often it is about what is "accomplished" rather than what is "conveyed" as ethologists and behavioural ecologists put it (Owings and Morton 1998). When it is about what is "conveyed", however, it is not obvious that it is appropriate to talk about memes irrespective of the social learning mechanism involved. With appropriate caveats about units as discussed above, the meme concept can be useful when the mechanism involved is linguistic because information is obviously being transmitted. But what about other mechanisms? Psychologists have proposed a variety of mechanisms that could be involved in social learning but we will consider only two other broad categories (for recent reviews see Zentall 2006, and the articles in Hurley and Chater 2005). Before doing so it will be useful to point out that, just as will be discussed in the next chapter about individual learning, we assume here that all social learning, including learning in animals, is cognitive – not in any sense necessarily related to consciousness, but in the sense that it creates psychological/neurophysiological programmes in the minds/brains of individuals.

Now consider a case of social learning taking place by individual learning mechanisms – by instrumental or operant conditioning. An episode of social learning between ego and alter with the direction ego to alter may actually be rather complicated. First, it may be in the (reinforcement) interest of ego to shape alter's behaviour to be similar to its own, in which case the interaction could be neutral or predatory with respect to alter. (Of course, it could accidentally be of benefit to alter as well but in that case it would be in ego's interest to foist at least some of the costs onto alter by making the signals "quieter", forcing ego to work harder to receive them and even for alter to be willing to pay some of the costs, thus transforming the situation into that of three below.) Secondly, it may be in the (reinforcement) interest of alter to have its behaviour shaped to be similar to ego's in which case the interaction could be neutral or parasitic with respect to ego (with a caveat analogous to that above). Or thirdly, it may be in both, i.e. in their mutual, interests. For example, the first was the implicit assumption in the original theory of social learning by individual learning mechanisms – the theory of matched-dependent learning (Miller and Dollard 1941, 1962) in which ego reinforces alter's behaviour for matching its own. On the other hand, the second is the general kind of assumption in Boyd and Richerson's theory of the evolution of cultural transmission according to which learning from others is favoured if learning individually is error prone or expensive, and environments are neither too variable (which would favour individual learning) nor too stable (which would favour genetic transmission) (Boyd and Richerson 1985, Richerson and Boyd 2005: chapter 4). Notice that, in the first kind of case, ego must learn individually how to emit signals (here cues which induce the expression of some particular behaviour in alter, warnings which suppress others, and/or rewards and punishments which select among alter's behaviours), in order to *shape* (as the psychologists call it) alter's knowledge and behaviour to be similar to its own. Because ego knows how and is able to do something does not mean that it knows how to shape another to know and do similarly. In the second kind of case on the other hand, alter must learn individually how to have its behaviour shaped by ego – i.e. to receive and correctly interpret signals (here cues, warnings, rewards and punishments) emitted by ego in order to have its behaviour rendered similar to ego's. In both cases, "signals" are signals in the ordinary language sense only to one, not both, of the parties. In the third kind of case, in which behaving similarly is in their mutual interests, they are signals to both and both are learning.

Given the assumption that all learning, including the social, is cognitive, in all three cases information has been created in one or more brains. The important point, however, is that in the first case information has been created in the brain of ego only; in the second in the brain of alter only; and in the third, information has been created in both, but it is *completely different information*. There would be no reason to believe that alter who has been shaped socially by such a mechanism would thereby be capable of shaping another in turn. Alter would have to learn individually to do so. In essence, these are social psychological and not sociocultural processes. There is no meaningful sense in which information has been transmitted. Memetic language to describe them would therefore be inappropriate.

While social learning by linguistic instruction clearly transmits information and social learning by individual learning mechanisms clearly does not, the case of social learning by observation (in any sensory modality), "learning to do by seeing it done", variously called observational learning, imitation, or true imitation, the situation is less clear. As noted in Section 2.3, the possibility or even the probability of imitation arises when the response is not in the previous repertoire of the learner, when the learner just observes not performs in the learning situation (and hence there is no possibility of reinforcement for performing at least), and when a long time lag exists between observation and performance (Bandura's original criteria e.g. 1977:36–37). According to Zentall's review (2006), the evidence for imitation is rather good for birds and primates. However, most ethologists and behavioural ecologists would at least extend the latter to mammals in general (including Cetaceans for example, Rendell and Whitehead 2001), and even to fish and insects in some cases – for example when the behaviour involved is female mate choice (for an overview see Dugatkin 2000). The copying of mate choice among females is a long recognized feature of lek mating systems in some species. In some sense the explanation of learning by observation would have to include what Bandura called a "covert symbolic process" (1977:36–37), essentially a process that Hodgson and Knudsen (2006) see as requiring an analogue of "back", i.e. reverse, translation. Such a process from a cultural evolutionary perspective would commonly be described as Lamarckian in the broad sense (i.e. the inheritance of acquired modifications, not necessarily preferentially of acquired adaptations – think teenagers). An organism might "reverse translate" the actions of another into cognitive descriptions of the behaviour, with the latter then guiding their own actions. In humans, this reverse translation could be into language, whether silent or overt.

It is likely that over the coming years we will be hearing a lot more about this form of learning and its neurophysiological base. The last two Grawemeyer awards – the most prestigious (and lucrative) awards in psychology were given for relevant work. As noted in Chapter 2, in 2008 the prize was given to the "social cognitivist" Albert Bandura. As also explained there, he first showed decades ago, stemming in part from his research on the effects of television violence on children, that observational learning could not be reduced to operant conditioning (for an overview see Bandura 1971, 2007; 1977). The year before, in 2007, the award was given to three Italian researchers from the University of Parma – G. Rizzolatti, V. Gallese and L. Fogassi – for their discovery of "mirror neurons" in the brains of macaque monkeys, neurons which "mirror" the actions of another by being active both when an action like grasping is observed and when it is performed (see Gallese *et al.* 1996; Rizzolatti *et al.* 1996). A similar phenomenon has also been found in male swamp sparrows who learn their song by hearing it sung (Prather *et al.* 2008). One might be tempted to argue that in the activity of mirror neurons, Aunger's (2002) prediction of the existence of a neurophysiological "electric meme" has been verified.

5.7   THE IDEAL AND THE MATERIAL

As we have seen, biological evolutionary theory can and has been formulated in ideal (genetical) terms as in population and ecological genetics or in material (phenotypic) terms as in most evolutionary ecology. The direction of the change has varied historically. Darwin and the field naturalists of his time formulated theories about organisms and their properties, not genes. Population and ecological genetics came later in the twentieth century but modern evolutionary ecology reversed the direction again, beginning in the 1960s and 1970s. The move in systematics in recent years on the other hand has been the reverse from that which created evolutionary ecology. Historically, classification was based on phenotypes rather than genotypes because those were the data available. However, now that we have entered the era of mass sequencing of the genomes of many species, despite our lack of knowledge of what most of these genetic sequences do, if anything, systematic research is increasingly based on genomic data. While neither the ideal nor the material approach is entirely satisfactory taken alone, pragmatically, both have obviously been useful. Moreover, we shall see in Section 7.2 how a definition of evolution by natural selection can be formulated in a way that includes both.

In the social sciences, whether to emphasize ideas or the material basis of existence has been a dilemma throughout their histories. Even linguistics, obviously focused on symbol systems, has theorists who emphasize the "pragmatics" that go beyond symbolic reference. As a practical matter, anthropology has always been divided between cultural anthropologists and archaeologists – those who study culture, commonly defined in ideal terms, versus those who study the material remnants of past human societies (albeit many of the former include material culture in their field of interest and many of the latter interpret archaeological artifacts largely in terms of their symbolic content). Sociologists have very explicitly disagreed on this matter. The content of the law of the three stages proposed by Auguste Comte, the founder of sociology, was explicitly idealist. What changed in the transition of societies, institutions and even individuals from the theological to the metaphysical to the positivistic stage, according to Comte (1855, 1974), was the nature of dominant ideas. Explanations in the three stages were proposed to be based on fictitious first and final causes, abstract forces, and positive (i.e. scientific) laws of succession and resemblance respectively.

Karl Marx held a quite different view. In Marx's materialist conception of history, the economic base of society included the "forces of production" (basically the prevailing technology) and the "social relations of production" (how people relate to the prevailing technology, for example as owners and non-owners of factories in capitalist societies). This was the material basis on which society is built. Everything else including the "legal, political, religious, artistic or philosophic – in short ideological" (1859, 1959), Marx viewed as derivative. They constituted the "superstructure" which rested on the economic base. The sociologist Max Weber tried to reconcile such extremes. He acknowledged the importance for the origin of capitalism of all the material factors like the importation of precious metals from America signalled out by Marx. However, he maintained in his studies of the world's major religions including his classic *The Protestant Ethic and the Spirit of Capitalism* (1904–5, 1958), that such factors had been present at other times and places without the same result. The "additional necessary factor" in the origin of capitalism in Europe, he believed, was a religious doctrine, protestantism, specifically Calvinism. The belief in a "calling", even to secular vocations, committed them to work hard, and their belief in predestination (through its effect on anxiety about being among the elect) made them resist wasting the results of their labour on "wine, women and song" so to speak. The result of the addition of these religious beliefs to the right material conditions according to Weber was capitalism.

Among the sciences of the human which begin with individuals rather than groups, theorists have disagreed but sometimes combine both. According to economists, people have ideas in their heads – about what they want (utility functions) and what they think they know (information). However, in conjunction with circumstances (and sometimes the choices of others), these ideas result in behavioural choices which in turn have material consequences for profits, wages, rents, etc. The problem has been quite difficult for psychologists who have tended to be divided between behaviourists and cognitivists, although many, like the cognitive behaviourist, Edward Chance Tolman (1949), combined both.

Where inter-generational transmission has been considered, the ideal–material dilemma has been acute as well. For example, sociologists have accumulated much evidence for the intergenerational transmission of "SES" (socioeconomic status) in modern societies. Moreover, the basis of the correlation is largely education – highly educated parents, particularly traditionally fathers, achieve a high occupational status and their children (traditionally sons) similarly achieve a high education and occupational status. But that does not answer the ultimate question of the mechanism. How much do children of high-status parents achieve their educational status through the material resources parents are able to bequeath to them and how much through the cultural transmission of norms and values such as the value of educational achievement?

It is not surprising then, given the history of evolutionary biology and the history of the social sciences themselves, that the ideal–material dilemma has persisted in Darwinian sociocultural evolution. We noted earlier even Dawkins' initial confusion on this question and his later clarification. Those who utilize the concept of memes have tended to settle on the ideal – to many, a meme is essentially an *idea* that spreads. Most cultural evolutionists, even while avoiding the "meme" meme, agree. Recall that according to Durham, culture is first and foremost an "*ideational* phenomenon", a "*conceptual* reality" – which is socially transmitted, symbolically encoded, systematically organized, and shared because of its social history (1991: Introduction, emphasis added). Or to quote Mesoudi *et al.* again (2004, 2006), paraphrasing Boyd and Richerson (1985) and Richerson and Boyd (2005), the subject matter of cultural evolution is *information* – "information such as knowledge, beliefs, and values that is *inherited* through social learning and *expressed* in behavior and artifacts" (emphases added). Moreover, Boyd and Richerson (1985: 70 ff) employed a quantitative genetics-type model, including the additivity assumption as a model for the effects of cultural transmission. Despite the majority view, many practising cultural evolutionists mirror evolutionary ecologists in taking as their subject matter the *material* manifestation in behaviour and artifacts of culturally transmitted ideas. George Basalla, for example, is quite definite in his view on this. As an historian and evolutionary theorist of technology, he views the artifacts themselves and their properties as his subject matter, not, for example, mental plans for their production (1988: introduction).

5.8 SUMMARY AND IMPLICATIONS

The meme concept was generally not enthusiastically received in academic circles, albeit the reception among those interested in

Darwinian-style theories of evolution in the social sciences was more mixed. Beyond the sociology of its invention and reception, objections to the meme concept that there are discrete units of symbolically encoded biological information which evolve – genes, but not of cultural information – memes, are not persuasive as was discussed. Both genetically and linguistically encoded information include units of structure which are discrete and of function which are symbolic, but because recombination can and does take place structurally within function units, in neither case do the two necessarily coincide. Hence the relationship between structure and function can be one to many and many to one (which is not to deny the capacity for adaptive evolution if effects combine additively). Despite the complexity, like the gene concept historically, the meme concept has been shown by Pocklington and Best (1997), Best and Pocklington (1999), Ritt (2004) and Jan (2003, 2007) for example to be capable of doing a useful job of scientific work. Although popular usage differs somewhat from the academic and a great variety of other terms are used in a variety of social science disciplines, the "meme" term may be particularly useful in interdisciplinary discourse. With respect to cultural inheritance, memetic talk may be most appropriate when social learning is by linguistic instruction and least appropriate when it is by individual learning mechanisms. The situation is less clear when social learning is by observation or true imitation, although the discovery of "mirror neurons" may eventually shed some light on this.

There is no reason to expect, nor should there be, the banishment of either the ideal or material perspective. Ideas may spread, but we all know that, like genes, their behavioural consequences can be very complicated indeed. Given that both ideal and material approaches have been useful at various times in the history of both biology and of the social sciences, and given that the meme concept in particular appears no more problematic than the gene concept, the most reasonable conclusion to be drawn seems to be "memes if useful but not necessarily memes".

6

Micro and macro I: the problem of agency

6.1 THE PROBLEM OF AGENCY

The problem of the relationship between individuals and societies/ cultures is an old one. Do the experiences and choices of individuals make societies and cultures or do societies and cultures structure and programme the behaviour of individuals? This dilemma was at the heart of the founding of the social sciences in the nineteenth century with economics, largely an English product, choosing the former and sociology and anthropology, largely products of continental Europe, choosing the latter. In the last quarter of the twentieth century, the prolific English sociologist Anthony Giddens (1976, 1979, 1984) dubbed this the problem of "agency" versus "structure" and the terminology has stuck – migrating beyond sociology into other social science disciplines such as institutional economics for example (Hodgson 2004).

The reader will notice, however, that this chapter is titled "The problem of agency" rather than that of agency vis à vis social structure/culture. We already saw in Section 2.3 on history how Bandura showed that sociocultural processes in the form of social learning cannot be reduced to the psychological in the form of individual learning. We also saw in Section 5.6 on memes how they cannot be reduced to social psychological ones in the form of interacting individual learning either. The problem of agency, i.e. of individual action, needs to be addressed in its own right, however. That is because too often in many of the social sciences today it is implicitly assumed to be more or less a matter of free will. The concept of free will has no scientific credibility. It is not a theory. Because it can explain everything, it can explain nothing and is in fact an excuse not to seek an explanation. Moreover, human psychology is not a black box and should not be viewed as such by social scientists.

The two most well-developed theories of agency or individual action apart from particular genetic and sociocultural influences are considered in this chapter. These are the most well-developed theories in the sense that each possesses in spades one of the hallmarks of a very successful science – in the one case a strong tradition of empirical, in fact, experimental support, and in the other a rigorously derived deductive theoretical structure. Notably, the credibility of both is buttressed by their real-world applicability – the one constituting the foundations of applied behavioural analysis, the other of microeconomics. The relationship between the first "pragmatic" perspective (learning theory) and the second "utilitarian" perspective (rational choice theory) has been much discussed (e.g. Herrnstein 1990), but ultimately we will suggest that they are complementary. The latter is needed to explain origins and the former to explain the maintenance of stability or the direction of change.

6.2 BEHAVIOURISM, TELEOLOGY AND TELEONOMY

Behaviourist psychology in general, and B. F. Skinner in particular, has a bad reputation, some of it deserved, among other kinds of scientists. The reasons for this are many but they include the following. Skinner's (1948a) utopian novel about a world in which everyone's behaviour is controlled by operant conditioning (specifically positive reinforcement) struck most outsiders rather as a 1984ish dystopia. The rejection of a role for theory and theoretical constructs or hypothetical entities in science was met with incredulity by historians and philosophers of science. Behaviourist psychology isolated itself from even closely neighbouring disciplines. For example, it was held that behaviour analysts had nothing in common with neurophysiologists who seek to understand the neurophysiological basis of learning; with cognitive psychologists who want to understand how animals and people acquire, process, store and utilize information as well as how they behave; or with social psychologists who want to understand how individuals learn socially as part of cultural communities as well as through their own unique individual experiences. Enemies were made even further afield – of evolutionists and geneticists by ignoring biologically evolved and inherited species-specific capacities for and constraints on learning and of linguists by claiming that all that is necessary to understand human language are the principles of operant conditioning. They engaged in generally sectarian behaviour – isolating themselves even from other psychologists studying learning in their own association, the "Society for the Experimental Analysis of Behavior" and rejecting the traditional disciplinary label of psychologist and of their field as learning in favour of "behaviour analysts" studying "operant behaviour". Partly as a result of the latter, the experimental study of learning degenerated into a fratricidal war between those emphasizing classical, type one or respondent conditioning (Pavlov's original S1 S2 R model of learning by association) and those emphasizing instrumental, type two or operant conditioning (Thorndike's original S1 R S2 model of learning by reinforcement and punishment) to the point that balanced texts in the field became rarer than either of the ideologically committed alternatives for a time. It is a sad story which partially accounts for the fact that behaviourism and the study of learning in general, beginning in the mid-1960s or so, came to be reduced from its former dominance in psychology departments in North America to a small minority. Despite, or perhaps because of, this woeful tale, many of the best behaviourists learned some lessons and, in the past two decades, while rightfully maintaining the principles that a long and successful experimental tradition discovered, there has at the same time been a flowering of engagement of those so inclined with others. Hence for example we have seen the experimental methods behaviourists developed become some of the most important tools in the study of neurophysiology and behavioural economics and many conversations taking place such as those between Howard Rachlin and Richard Herrnstein with economists and Sigrid Glenn with evolutionists and immunologists for example.

The fundamental problem that the theory of instrumental or operant conditioning (and indeed all "selectionist" or "consequentialist" theories) solve is the problem of teleology. This can be illustrated with a very simple example. We place a hungry rat in the start box at the bottom of a T-shaped maze with a small amount of food in the left arm of the T at the top. After a few minutes of sniffing around in the long arm of the maze it will either turn left or turn right at the top. Whether sooner or later, however, it will turn left and obtain the food. We then remove the rat, replace the food, and run the experiment on the same rat a few more times in succession. After a few trials, it will run straight down and turn left immediately – the time taken to do so generally decreasing on subsequent trials. Now, we look at the most recent trial and ask the question, "Why did the rat turn left?" The common sense answer is, "The rat turned left in order to get food". But wait a minute. In this trial the rat got the food at time T2 but it turned left at time T1. Something that happened later at time T2 cannot be the cause of something that happened earlier at time T1. Implying that causality can work backwards in time is a mistake in logic called "teleology". In fact, it is not the food received in this recent trial that explains its having turned left, it is because food was received as a consequence of turning left on previous trials that explains why it turned left this time. It is **2** below that explains **3**, not **4**.

1. Turns left   **2.** Receives food   →   **3.** Turns left   **4.** Receives food

Behaviours (here turning left) which have the consequence of or are followed by reinforcement (here food) become relatively more probable (here relative to turning right), and with some complications, those followed by or having the consequence of punishment (such as a mild electric shock) become relatively less probable. This "law of effect" was the insight of Edward L. Thorndike (1911, 2000) who originated the very productive tradition in the experimental study of individual learning by instrumental or operant conditioning which was discussed briefly in Chapter 2 (there in the context of pointing out that animals and people also learn socially).

Note that we could have partially corrected our original mistake by expressing ourselves more carefully and answering instead, "The rat turned left because it *expected* to get food there". That way the cause, now the expectation, would have been placed in front of the effect, getting food, where it belongs. But that would not have been the final answer either. We would have then had to go on and ask, "Why did the rat expect to get food by turning left?". It expected to get food by turning left because of its past history of reinforcement. It was the fact that turning left on previous trials had the consequence of or was followed by the rat obtaining food which created that expectation as opposed to others (the expectation intervening between **2** and **3** in the sequence above). There are good reasons (discussed below) for believing that all learning, even in organisms with quite simple nervous systems, is "cognitive" in this sense of creating expectations, i.e. information in brains/minds (which has nothing to do with consciousness in the sense of self-awareness, which is a different question).

It is now widely recognized that teleological statements - those which make reference to future states as causes - on their own, are not so much wrong as meaningless. They literally do not make sense (for a dissenting view see Walsh 2008). On the other hand, often they are elliptical statements about selection processes. In a biological context, we say the function of the heart is to circulate blood but what we really mean is that once upon a time in a population of organisms there were individuals with hearts that did, and those with hearts that did not, circulate blood (or more realistically, which circulated it with varying degrees of efficiency). Those organisms with hearts that circulated blood efficiently survived and reproduced at higher rates than did those with hearts that did not, and because of the principle of heredity, that offspring resemble their parents, members of the relevant population today have hearts that efficiently circulate blood. The point is relevant as well of course to the long history of debates over functional explanations in anthropology and sociology. Functional statements about institutions such as political, economic or educational institutions, for example, would make sense only in the context of an implied selection process. To say that the function of political institutions is to reconcile conflicting interests would only make sense, whether or not they are true, in the context of claims about competing societies which differed in viability because of the presence or absence of political systems which performed in this way. In a classic article, Pittendrigh (1958) suggested that we substitute the terms *teleonomy* and *teleonomic* for teleology and teleological in such contexts to distinguish statements made in the context of some selection process which do make sense from those which are not and therefore do not. Selection processes which have been analysed in this way include not only biological evolution, but also some processes which take place within individuals including individual learning but also the adaptive immune response, and of course sociocultural evolution.

6.3   INDIVIDUAL LEARNING AS
A SELECTION PROCESS

To return to the specific case of individual learning, there is a long but only sporadically connected history of pointing out the similarities between the processes of biological evolution by natural selection and psychological/neurophysiological learning by reinforcement and punishment (see, for example, Pringle 1951; Campbell 1960; Russell 1962; Skinner 1966; Jerne 1967; Gilbert 1970, 1972; Staddon and Simmelhag 1971; Staddon 1975; Edelman 1987; Vaughan and Herrnstein 1987; Glenn 1991; Rachlin 1991; Glenn, Ellis and Greenspoon 1992; Glenn

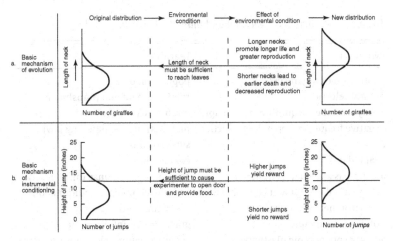

Figure 2. The basic mechanism of evolution compared with the basic mechanism of instrumental conditioning. The two examples used are (a) the natural selection of long-necked giraffes and (b) a jumping cat experiment. Reproduced with permission from p. 144, Howard Rachlin, *Introduction to Modern Behaviorism*, 3rd edn. 1991. New York: W. H. Freeman and Company.

and Field 1994; Plotkin 1994; Cziko 1995; Glenn and Madden 1995; Hull, Langman and Glenn 2001. Rachlin (1991:144) illustrated the basic similarity by comparing the statistical shifts which take place in evolution with those which take place in learning (specifically, by rewarding a cat with food each time its jump exceeds a particular height, see Figure 2). In fact, the similarities go well beyond this basic one (Blute 2001b; Hull, Langman and Glenn 2001.) Some prefer to view the two processes as analogous; others prefer to view them as different tokens of the same general type – selection processes. On whichever view, consider the following points of correspondence (which for convenience are summarized in Table 4).

6.3.1 The Hardy–Weinberg equilibrium and matching, negative frequency-dependent selection and melioration

Evolutionists modestly call their laws "principles". Among these the most fundamental is the Hardy–Weinberg equilibrium principle according to which the relative frequencies of alternative alleles (proportions of different versions of a gene in a population) remain constant unless some force like natural selection acts to change them. Learning

Table 4. *Comparison of evolution and individual learning*

Evolution	Learning
population of organisms	population of acts
natural selection	reinforcement and punishment
Hardy–Weinberg equilibrium principle	(probability) matching
negative frequency-dependent selection	melioration (the matching law)
drift	superstitions
adapted or fit	competent
anagenesis and cladogenesis	shaping and response differentiation
organism's development conditional on environmental induction	act's development conditional on discriminative stimuli
Von Baer's principle	gradient of reinforcement
adaptive phenotypic plasticity; rule-based genetic programme or algorithm	cognitive nature of learning; rule-based psychological/neurophysiological programme or algorithm
a single evolutionary process	a single learning process (see Figure 3)

theorists too speak of "matching" in the sense that relative frequencies or rates of responding tend to match relative frequencies or rates of reinforcement and hence the former should not change unless the latter does (sometimes called "probability matching"). However, according to Herrnstein, who introduced the "matching law" (1970, 1990, 1997), what is actually observed is "melioration" and is akin to the principle of diminishing returns in economics or negative frequency-dependent selection in evolution. "This empirical law states that choices are allocated such that the accumulated rewards harvested from an alternative, divided by the number of times it has been chosen, is equal for all alternatives" (Loewenstein and Seung 2006:15, 224). In whatever form (frequencies, rates, or profitabilities so to speak), such principles constitute a base line against which to measure change and in that sense resemble Newton's first law of motion.

6.3.2   Drift and superstitions; adapted or fit and competent

We carefully said above that acts which have the consequence *or* are followed by reinforcement become relatively more probable. That is because acts can be "contingently" or "functionally" (Glenn, Ellis and Greenspoon 1992), i.e. *causally* connected with reinforcement (as when a pigeon receives food for pecking a disk which has been mechanically set to deliver food only when pecked). Alternatively, they may be *accidentally* so associated (as when reinforcement is delivered at fixed temporal intervals independently of behaviour and whatever idiosyncratic act which happened to precede it becomes more probable) – a phenomenon that Skinner (1948b) discovered and famously dubbed as a "superstition". Evolutionists call the equivalent of the former phenomenon "selection" attributable to "fitness" or "adaptedness" and the second "drift". In a large population of some black and some white rabbits living in a snow-covered environment, if wolves can see black rabbits better against the snow, then because of selective predation, all other things being equal, we would expect the population of rabbits to shift against the black and in favour of the white. But if the environment were not snow-covered, or if wolves had no visual ability to discriminate among rabbits by colour, in a small population we might nevertheless observe a shift in one or the other direction. This is no different than a coin tossing experiment. We expect an equal number of heads and tails to turn up in a very large number of tosses but not in every short sequence of such. Just as in a finite biological population, where the characteristics of organisms can "drift" from one to another in the absence of adaptive evolution, in a superstition experiment, acts are commonly observed to drift from one to another in the absence of causally contingent reinforcement. (It is worth noting that learning theory has no term equivalent to fit or adapted and it is sometimes useful to have such a term – "competent" can be useful here. A "competent" act would be one which has the requisite properties in a given situation to cause reinforcement.)

6.3.3 Anagenesis and shaping; cladogenesis and response differentiation

Both evolution and learning can take place in two fundamentally different patterns. Evolution can take place with innovations including mutations and recombinations being followed by selection resulting in gradual change through time within a population. While there may be no single point at which we can draw a firm line, these kinds of events can eventually result in such a large amount of change that if we were able to look at the population at one point in time and jump to a much later time, the change would be so great that some, specifically "evolutionary taxonomists" as discussed in Chapter 2, would call the later population members of a different species. This form of evolution "in a straight line" so to speak is called "anagenesis". The corresponding pattern in learning is called "shaping". An example would be teaching a dog to fetch a stick. As variation appears, by reinforcing successive approximations of the target behaviour – looking in the direction the stick is thrown, turning its shoulders in that direction, taking a couple of steps, running to the stick, picking it up, heading back with it, returning it, etc. – a new response class "stick fetching" evolves out of the old "orienting response". The other pattern in evolution, the branching of one species into two emphasized by cladists as discussed in Chapter 2, is called "cladogenesis". The corresponding pattern in learning is called "response differentiation". Both left and right turns may "evolve" among a rat's behaviours in a T-shaped maze out of sniffing around in the long arm if both alternatives are reinforced under different signalling conditions.

6.3.4   Development and environmental induction/ discriminative stimuli

No organism or act is an instantaneous event – even the simplest have a development, i.e. a life cycle, and the entire cycle is affected by natural selection or reinforcement. Moreover, the properties of such cycles are dependent, not solely on the history of natural selection or reinforcement, but also on the environmental circumstances encountered during each development. One may inherit a genetic tendency to be tall, but that height will only be realized "phenotypically" if inductive influences stemming from the environment during development such as nutrition are the same or similar to those under which ancestors developed and were selected. Similarly, in learning, reinforcement is said to bring behaviour under the control of discriminative (signalling) stimuli. A child new to school learns to raise his or her hand before speaking out but will not normally thereby raise his or her hand before speaking at the dinner table. In short, we could say that just as evolution is change in the inductive control of organismic development by the ecological environment (Van Valen 1973; Blute 2007, 2008a) learning is change in the discriminative control of act development by the environment (this definition will be expanded on in Section 7.2).

6.3.5   Von Baer's principle and the gradient of reinforcement

The nineteenth-century German embryologist Karl Ernst von Baer observed that members of different but related groups are more similar earlier in development than they are later. Von Baer was not an evolutionist but his principle implies that more change has taken place in evolution later in development than earlier. This is not a selection principle, but a constraint one. Adaptation presumably is just as important and therefore selection should be similarly intense at all stages of development. However, what happens later in development is in part determined by what happens earlier. A mutation which has substantial effects on an embryo early in development would, on average, be less likely to be adaptive than one whose major effects appear later because of the complex ramifications the former would have. In modern times, Wimsatt and Schank (1988) dubbed this principle "generative entrenchment". It was long viewed as an approximately true generalization although exceptions were known.

Clark Hull, one of the great systematizers of learning theory in the first half of the twentieth century, explored the corresponding phenomenon in learning, the principle of which had first been noted by Thorndike. A rat learning a maze with a series of choice points tends to eliminate mistakes in a backward direction and when reinforcement is scheduled for the first response after a fixed interval, responding accelerates through the interval. This "gradient of reinforcement" is commonly hypothesized to exist because activities closest in time to reinforcement are learned most easily. Apparently rats even "think" backwards when recalling something (Foster and Wilson, 2006)!

An interesting twist on this "more change later than earlier" phenomenon in biology arose in recent decades when it was pointed out that it does not apply to the very earliest stages (see discussion in Raff 1996: 190–210; Hall 1997). Elinson (1987:2) illustrated this in a particularly striking way by adding drawings of the three earliest stages (egg, blastula and gastrula) of three different vertebrates (salamander, chick and human) to the embryonic stages used by Haeckel to illustrate his version of the principle. The drawings show that the differences between groups are large at these very early stages and then decline before they increase again as commonly illustrated. This has been termed an egg-timer or hourglass pattern and the point of minimum difference "phylotypic", in vertebrates the "pharyngula", or viewed as a particular pattern of gene expression at that stage in animals, the "zootype". Elinson suggested that major changes may take place early because of major ecological shifts – for example to a terrestrial way of life necessitating large yolky eggs or maternally supplied nutrition although other differences such as those among different orders of amphibians or species of frogs are less readily explicable. Raff suggested a possible developmental constraints explanation. A single-module, early-stage embryo is flexible. As it begins to become modularized, the modules interact with each other reducing flexibility. But later, more numerous fully formed modules are more independent of each other increasing flexibility again.

A possible evolutionary explanation might begin with accepting that the original point is logically unassailable – more change should be expected later than earlier. If that is accepted, but empirically more change than expected is observed very early, then one might conclude logically that what has been thought to be earliest stage(s) must in fact *be* later. This could be the case if what was traditionally viewed as the earliest stage(s) of development are in large measure the result of maternal effects and their ramifications. (This might then raise the question of why change there is not even greater than it is. However, it is only in semelparous organisms that maternal effects on offspring are selected for solely at the very end of the maternal life cycle.) Consistent with a maternal effects interpretation of the "too much" early change (not incompatible with an ecological explanation), the phylotypic stage or point of minimum change in the life cycle should then be seen as the point in the life cycle of the actual historical origin of the taxon involved. Moreover this does lead to a broad prediction – all other things being equal, the more parental influence, commonly maternal care, the more "too much" early change should be observed.

There are a number of phenomena or principles known in evolution but not in learning, and vice versa, that could be worth investigating. For example, a large part of the study of learning has involved studying the effects of various schedules of reinforcement – a phenomenon unknown in research on evolution but which could potentially be investigated with the kinds of "evolution in a test tube" experiments sometimes utilized there with organisms such as *E.coli*. In the opposite direction, although the gradient of reinforcement remains a phenomenon of interest in learning, the hourglass pattern discovered in evolution has not been observed there to my knowledge. However, I doubt it ever will be if the explanation suggested above is correct. That is because, given that the data are generated in laboratory experiments, we *know* without doubt what the historical origin of those sequences are.

6.3.6   Adaptive phenotypic plasticity and the "cognitive" nature of learning – rule-based programmes or algorithms

It is often observed that organisms develop, within limits, in an equifinal fashion. Some plants develop one phenotype if a seed develops on land and a completely different one if it develops in water, each adapted to the circumstances which induce it. Genetic programmes or algorithms can apparently include sets of rules which constitute alternative instructions for achieving the same result under different circumstances – in some cases presumably because a variety of environments have been present in the ancestral past and have therefore been circumstances that the genotype has evolved to be able to respond adaptively to. In other cases, their availability may be a purely fortuitous pre-adaptation. This phenomenon of equifinality – originally called regulative development by embryologists, canalization by Waddington in the 1940s, and adaptive phenotypic plasticity by modern evolutionists (e.g. Pigliucci 2001) – was discovered in development by Hans Driesch (1914) around the turn of the last century. In one early experiment, Driesch removed one of two cells of a sea urchin embryo at the two-celled stage and, eventually, a whole but small embryo developed anyway! Later research has revealed many such phenomena. For example, with the early limb buds of a salamander:

> If half a limb bud is destroyed, the remaining half gives rise to a
> completely normal limb. If a limb bud is slit vertically into two or more
> segments, while remaining an integral part of the embryo, and the parts
> are prevented from fusing again by inserting a bit of membrane between
> them, each may develop into a complete limb. If two limb buds are
> combined in harmonious orientation with regard to their axes, a single
> limb develops that is large at first but soon is regulated to normal size
> (Berrill, 1976:309).

And of course, much of such adaptive phenotypic plasticity is observed not only in development but also in physiology and particularly behaviour. Such (limited) equifinality once supplied motivation to those who embraced vitalism as a philosophy in the life sciences. Driesch himself was such an advocate, claiming that only the presence of an "entelechy" or life force could account for such apparently purposeful behaviour. Today, however, we understand that such phenomena are largely attributable to a complex hierarchy of control of gene expression. Genetic algorithms, like computer programs, include rules – whole routines, subroutines, and sub-subroutines, etc. – whose execution is controlled by a variety of stimuli originating both internally and externally. Each routine and subroutine is commonly adaptive with respect to the conditions which induce it. Despite this, it remains the case that new variations in programmes do not arise preferentially in an adaptive direction which would return us to naive teleology again.

Demonstrations of the corresponding "cognitive" nature of learning go back to experiments in place learning performed by the cognitive
behaviourist Edward Chance Tolman in the 1940s (e.g. Tolman, Ritchie
and Kalish 1946). For example, in a four-arm maze, if you first teach rats
starting from the south to get food in the west box, then test them
starting from the north box and they still get the food, then they obviously have not learned only a behaviour (i.e. turning left which would
have led them to go east when coming from the north), but instead they
have learned to know, or more accurately to expect, where the food will
be. This can be most parsimoniously explained with the concept of a
programme or algorithm – what was increased in frequency by the
reward in learning was not a behaviour, turning left, but the execution
of a psychological/neurophysiological programme or algorithm which
minimally contained the rules "If coming from the south, go left; if
coming from the north go right" ("south" and "north" here being simply
markers for any discriminable features of the environment such as a
scratch on the floor or a light overhead). The full content of the programme did not become obvious, however, until the testing phase. (Of
course no programme includes all conceivable ways of achieving a result
which would again imply a naive teleology. For example, if we cemented
up the west arm for testing, the food would not be acquired. In that sense
there is no such thing as learning to achieve a goal without qualifica-
tion.) However, most behaviour in animals, even those with quite simple
nervous systems, let alone in people, probably involves the execution of
complex rule-based programmes including routines, subroutines and so
on. It is for this reason that I generally use the term "acts" rather than
"behaviours".

6.3.7   A single evolution and a single
learning process

Finally on individual learning, consider how the instrumental or
operant-type learning we have been discussing relates to other procedures yielding learning. Assume that "respondents" (behaviours
"elicited" by the environment and commonly studied by Pavlovian
associative procedures) and "operants" (behaviours freely "emitted"
and commonly studied by instrumental or operant procedures) are
not disjunct categories but that there is a continuum between them.
If we also then take into account (i) that acts develop; (ii) that they differ
in their "competence" in the circumstances; and (iii) that just as can
environmental influences on organismic phenotypes in evolution,

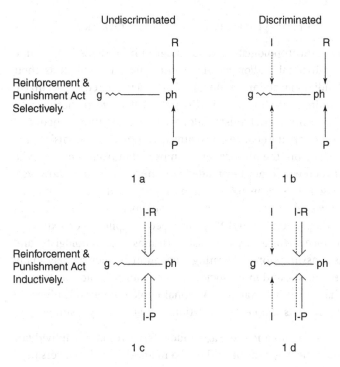

Figure 3. Relationship among learning procedures as aspects of an evolution-like process. Reproduced with permission from Marion Blute, A single-process learning theory, *Behavioral and Brain Sciences* **24**:530, 2001.

"forward-acting" stimuli in learning can affect the morphology and hence the competence of acts – then viewing learning as an evolution-like process can encompass under one roof, not only instrumental/operant conditioning, but all of the elementary procedures known to yield learning – a possibility that has for long been of interest (e.g. Blute 2001b; Donahoe and Vegas 2004; see also Figure 3;).

6.4 RATIONAL CHOICE AND GAME THEORY

Historically, anthropologists and sociologists have tended to assume learning (individual and/or social – commonly unspecified) as their theory of individual action – individuals are enculturated or socialized. (Social learning was discussed in Chapter 2. It is worth noting that individual learning mechanisms alone could constitute a theory of social interaction if the cues, warnings, rewards and punishments come not just from the physical environment, but from other individuals. That such would not be an adequate theory of culture, however, was discussed in Section 5.6 on memes and social learning mechanisms.) However, some traditional sociological theories such as exchange theory (e.g. Blau 1964) have been implicitly or explicitly rational choice theories. By contrast with most anthropologists' and sociologists' assumption of learning – economists, increasingly many political scientists, and more sociologists and anthropologists than in the past have been embracing a rational choice theory of individual action. Briefly in its simplest form rational choice theory assumes that:

(i)   scarcity obtains and there are trade-offs, i.e. resources individuals devote one way cannot be devoted in another and vice versa;

(ii)   individuals have goals, purposes, intentions, motives, preferences, utilities (the terminology hardly matters) – actually an ordered, stable set of such desired future states;

(iii) individuals have complete information, i.e. they know the conditions they face and therefore what the consequences of different courses of action will be; and

(iv)   as a consequence, they act so as to get what they want, "maximize their expected utility".

In short, individuals are egotistic, rational, utility-maximizers as it is commonly summarized or put more softly their behaviour is based on "beliefs, preferences and constraints" (Gintis 2007). Given such assumptions, a number of important theorems can be proven about how individuals are expected to behave. Among the most important is the marginal value theorem – that a rational actor will allocate resources among alternative courses of action in such a way that the "utility" or satisfaction received from the last (i.e. marginal) unit allocated to each is equal. Related to this is the principle of diminishing marginal utility. Typically, while the satisfaction received from successive units allocated to some particular course of action may increase initially (the second potato chip may taste even better than the first), eventually it tends to level off and even decline (how many movies or ice cream cones after all can one enjoy in succession!)

Like behaviourist psychology, the theory of rational choice has had more than its share of critics. According to some, despite the success of some market experiments pioneered by Vernon Smith in verifying familiar predictions about demand, supply and price, so much other falsifying evidence has been produced by experimental economists and cognitive psychologists that it should be viewed only as a normative theory about how rational actors *should* behave, not as a scientific theory about how they normally do. For example, it is difficult to believe that many preferences remain stable even from day to day let alone across the life course. A classic study showed that twenty-two major lottery winners were not happier than the same number of controls measured between one month and one and a half years after their big win (Brickman, Coates and Janoff-Bulman 1978). Apparently, the old adage "watch out what you wish for because you might get it" contains much truth.

According to others it can be considered a scientific theory, but only if significantly modified (e.g. the prospect theory of Kahneman and Tversky 1979 and Tversky and Kahneman 1992, which is said to more realistically model actual risk averse and risk seeking behaviour in people) or if utilities are independently measured. A fundamental issue is the problem of mistakes. Consider two possible courses of action A1 and A2 and two corresponding outcomes O1 and O2. An individual takes action A1. Is that because they preferred outcome O1? Perhaps so, but perhaps not. Perhaps they made a mistake – they may have preferred O2 but their knowledge of the circumstances and hence the consequences of A1 was incorrect. The problem of unintended consequences is ubiquitous in human affairs (e.g. see Clark 2005 on the role of French intendants in state–society differentiation).We may partially correct for incomplete information by assuming knowledge of the probabilities of outcomes only, but that does not solve the problem if expectations are systematically biased, which they commonly are according to cognitive psychologists (for an overview see Goldstein 2008: chapter 12). In an amusing exercise, that collective brain Wikipedia (2009) has assembled a list of 104 no less different kinds of cognitive biases that have been documented or at least proposed! We tend to attribute good outcomes to rational choices – we have a good driving record because we are careful, keep our job because we work hard, and are successful because we behave intelligently. We tend to attribute bad outcomes like an automobile accident, losing our job or being otherwise unsuccessful to accidents. But that is only for ourselves, relatives and friends. For others we tend to assume the reverse, attributing bad outcomes to choices – "it was their choice" – and good ones to accidents! A solution to the problem of mistakes is to independently measure utilities – for example by Von Neumann and Morgenstern's (1944) "standard reference gamble" or even by eliciting them verbally. That is not the approach most commonly taken, however. Instead, more detailed assumptions about goals or conditions are added from which eventually empirical predictions can be derived. According to still others, rational choice is a reasonable scientific theory but only on an aggregate level, a point of view that we will get to in Section 6.5 below.

Some other issues that have been identified can be handled by the theory. For example, objectives do not have to be confined to material income and wealth (albeit experimental economists stick closely to the use of money incentives in their experiments.) But power (the ability to demand compliance), status (the ability to command admiration and emulation) or anything else for that matter traditionally of interest to other kinds of social scientists can be the objective. Even the methodological individualism and egoistic assumptions can be dropped according to some – groups, organizations, nations, etc., can be rational actors (which is more or less the assumption of "realist" theories of international relations) and goals are not necessarily selfish but can be "other-regarding". Others, however, think that such moves negate the core of the theory.

In its narrowest form, rational choice theory deals only with interaction with the rest of the environment rather than with other competitors. A student choosing from among essay questions to answer on an exam based on her assessment of how difficult the question is and how well she knows the required material in each case is behaving rationally in the narrow sense. But if she makes her choice dependent upon what she thinks other students will choose, she is behaving strategically. For example, perhaps she thinks that the professor will get bored grading the same question over and over and, as a consequence, may impatiently mark that question harder. If so, then she might choose to answer the question she thinks the majority of other students are unlikely to choose. Most notably for students of politics and societies, the theory of rational choice in the narrow sense includes only competition and excludes all overt conflict. As Knight (1921, 1971) famously put it, it excludes "all preying of individuals upon each other" including "fraud or deceit, theft or brigandage" (quoted in Lindenberg 2001:638). "Strategic interaction" including overt conflict can be handled not by rational choice theory in the narrow sense, but by game theory. A significant problem, however, is that while commonly two-player zero-sum games are analytically tractable (chess-playing computer programs and all that), many-sided and non-zero-sum games commonly are not and hence much theoretical development depends upon the simulations of "agent-based models". In the non-economic social sciences, it is common to consider optimization and game theory as both parts of the same broad rational choice paradigm (e.g. Heckathorn 2001). In both the narrow or broad understanding, however, while the theories or models are built on the actions of individuals, they result in predictions of aggregate outcomes (like the famous theorem of the efficiency of perfectly competitive markets for example) which is why they are of interest to social scientists.

As a consequence of some of these issues and others, some non-economists tend to prefer much more expansive versions of the theory such as Lindenberg's (1990, 2001) RREEMM – "human beings are resourceful (search and learn); restricted (confronted with scarcity and choose); expecting (generate expectations); evaluating (have goals which lead to preferences); motivated (are motivated to achieve a condition which they value more highly than the one they are in); meaning (try to make action situations meaningful in terms of other elements of RREEM" (2001:663).

While intellectual historians have identified many under-appreciated forerunners of rational action theory in all the non-economic social sciences, there is no doubt that the modern pioneer was an economist, Gary S. Becker. In articles and books on topics such as discrimination, human capital, crime and the family, he pioneered a unified theoretical and empirical economic "method of analysis" as he put it which, while economic in origin, nevertheless departed sharply from traditional economic assumptions that people necessarily seek personal, material self-interest. As he put it in his Nobel lecture, on "the economic way of looking at behavior", "behavior is driven by a much richer set of values and preferences" (1993). Some other important modern landmarks in developing the theory by other kinds of social scientists themselves were Olson (1965, 1971) in political science who analysed the effect of group size on the possibility of achieving collective action, Posner (1972, 2007) in law on many legal topics, and Coleman (1990) in sociology who attempted to explain how effective social norms emerge from rational interaction among individuals among other things. The rational action approach to traditional political questions has been represented by its own journal since 1966, *Public Choice*, and to sociological questions by *Rationality and Society* since 1989 but has been significantly less represented in anthropology.

It is difficult not to think that, despite the sometimes challenging technical apparatus, rational choice theory is rather popular because it comes closest to our common sense explanations of behaviour. I went to the store because I wanted a loaf of bread; she went to medical school because she wanted to become a doctor and so on. Theorists of such often admit as much. "The rational choice approach to marriage and other behavior is in fact often consistent with the instinctive economics of the common person" (Becker 1993:396). Even in anthropology, where it has been less popular, it is implicit in much ethnographic research where an anthropologist approaches the apparently strange customs of an unfamiliar society and shows how they make sense after all. For example in his classic *Cannibals and Kings* (1977), Marvin Harris argued that the seemingly strange Indian sacred cow complex made sense because Indians derive more benefit from cows as draft animals and producers of dairy products and dung for fuel and fertilizer than they would from butchering them. Although Harris called his theoretical approach "cultural materialism" (sometimes called cultural ecology), in the final analysis, he typically appealed to the rationality of individuals. As Sanderson (2007:177) put it "Harris's whole concept of adaptation is actually informed by a thoroughgoing cost-benefit analysis" and he quotes Harris from his *Cultural Materialism: The Struggle for a Science of Culture* that cultural change "has (up to now at least) taken place through opportunistic changes that increase benefits and lower costs to individuals."

6.5   A BRIEF COMPARISON

Learning and rational choice can both be viewed as theories of choice in the sense of explaining outcomes among alternatives. Neither the "backward looking" learning theory nor the "forward looking" rational choice theory are teleological in the crude sense. That is obvious in the case of learning but rational choice too has its preferences and beliefs where they belong in front of the behaviour they are said to explain. They are similarly deficient in one sense that learning theory has no general theory of why a reinforcer is reinforcing and rational choice theory has no general theory of why preferences are as they are. Both tend to begin by alluding vaguely to biology – food, opportunities for sexual activity, etc., are rewarding or sought but learning theory has a definite advantage in three respects. First, it has an edge in explaining how the more varied rewarding states with which we are familiar apart from such biological basics are acquired – by secondary reinforcement. Events correlated or associated with reinforcement become reinforcing themselves. Secondly, it most definitely has an edge in explaining rather than assuming beliefs, i.e. expectations are a result of a past history of reinforcement. Thirdly, with their understanding of shaping and response differentiation, learning theory has a handle on the emergence of wholly new kinds of behaviours. Rational choice theory on the other hand has tended to be restricted to a single continuum or a fixed set of alternatives although that has changed somewhat with Romer's "new" or "endogenous" growth theory according to which new knowledge (new "recipes" or "rearrangements" of inputs) can have increasing returns to scale because they can have more than one use or be used by more than one person (for accessible introductions see Bailey 2001, Romer 2008).

Under the most restrictive assumptions, however, the predictions of the two theories tend to be similar. That is so much the case that, as we saw in Chapter 4, evolutionary ecologists with their own "consequentialist" theory of change by natural selection have nevertheless found it useful to borrow optimization and game theory models from rational choice theory to use in their own domain! It is also revealed in how commonly the two approaches are confounded in the non-economic social sciences. For example George Homans (1961, 1974) is often cited as a pioneer of rational choice theory in sociology. However, Homans himself thought he was applying learning theory at the time and in fact his "elementary forms" of social behaviour was a mixture of both so that his earliest propositions speak of rewards and stimulus control from learning theory while the later ones speak of choosing alternatives with the highest expected value from rational choice theory. Later, in reviewing a collection of essays of Coleman's, Homans expressed the view that "behavioral psychology is the more general theory from which rational choice may itself be derived" (1987:770). Whether the predictions of the two theories are exactly the same has been a matter of some debate and experimentation. Herrnstein's original research argued that optimization and his matching law (melioration) are not equivalents because the latter implies that behaviour in choice situations will shift towards higher value alternatives regardless of the long-term effect on overall reinforcement and can be followed in the articles collected in Herrnstein (1997). Two accessible treatments of the issue with somewhat different points of view are contained in Rachlin (1991:250–284) and Powell, Symbaluk and Macdonald (2005: chapter 10). The issue is important because, according to psychologists, it has implications for understanding addictive behaviour. As noted in the discussion of two versions of matching in Section 6.3 above, the difference corresponds in economic theory to competition without and with diminishing returns respectively, and in evolutionary theory to the distinction between ordinary ecological selection and social selection in the form of a negative frequency-dependent game respectively (Herrnstein 1997: chapter 10). As we saw in Chapter 4 with the density-dependence example, interestingly, according to evolutionists, either or even both together may obtain depending upon conditions. Similarly, surely our exam-taking student could be influenced by both the difficulty of the questions, her knowledge of the material *and* her expectations of what questions other students will choose to answer.

In any event, rational choice theory understood broadly assumes a prodigious capacity for calculation on the part of individuals – calculus for maximizing a continuous utility function, perhaps simulating for engaging in strategic interaction and so on. Given that and some of the other advantages learning theory has as mentioned above, a number of theorists have suggested that rationality tends to obtain, but only on an aggregate level. Hence for example, perfectly competitive markets may be efficient, not because individual firms are so much, but because "creative destruction" in Schumpeter's famous phrase obtains. Inefficient firms shrink, go out of business or are taken over, efficient ones grow, remain in business and take over the less efficient and so on – i.e. viability differences, selection and evolution obtain among firms. As noted in Chapter 2 a large literature has developed based on this view of organizations (e.g. most recently see Durand 2006).

According to others, rational action on the part of individuals can similarly be seen as an aggregate phenomenon. The brain after all is known to be a massive parallel processor. Acts compete within individuals so that just as adaptation is achieved in evolution, not because it is what is sought, but because natural selection controls what evolves, it is similarly the case that competence is achieved in learning, not because it is what is sought, but because reinforcement and punishment control what is learned. Essentially we optimize and play games because we *learn* to do so. This perspective was systematically developed by Fudenberg and Levine (1998) but has commonly been independently adopted by a number of sociologists (e.g. Abell 2000; Macy

2006). Much experimental economics currently is also preoccupied with the interaction between learning and reason. Either kind of aggregate view is more than a bit ironic because, in a game-theory context at least, it entails borrowing *back* from evolutionary biology, the modifications made there to make game theory suitable for populations (Maynard Smith 1982, Maynard Smith and Price 1973), theory which the biologists had borrowed from theories of individual action in the first place! As Abell (2000:241) put it, "If I care to pin-point any one analytic framework which might prove to be that most beneficial in the development of a genuine social theory, I would select evolutionary game theory. It models situations (i) with a strategic structure, (ii) which do not depend upon hyper-rational assumptions, and (iii) where dynamics select equilibrium outcomes".

Learning and rational choice are in many respects a classic "chicken and egg" problem. It would appear, however, that we should start with learning – individuals learn so they come to expect so they learn so they come to expect, etc. – right? Perhaps. On the other hand, let us return briefly to that rat in the T-maze with which we began. Consider its behaviour on the very first trial before it learned anything. First, it had some goal, purpose, intention, motive, preference, utility function (again the terminology hardly matters) right from the beginning. It was "motivated" as the psychologists say. It wanted food. The experimenter made sure of that by depriving it of food somewhat before the experiment began. Secondly, given that it did something on the very first trial – whether sniffing around in the long arm of the T, going left, going right, or whatever – it had some belief, some expectation about where food might be found *even if that expectation was wrong*. It would seem then that the "forward looking" rational choice or utilitarian perspective is right in origins but the "backward looking" learning or pragmatic perspective is right in explaining the maintenance of stability or the direction of change – in this case learning to turn left. In Chapter 8 on the evolution of complexity we will suggest that this dual perspective is generally required in understanding selection processes including the biological and the sociocultural.

6.6 SUMMARY AND CONCLUSIONS

This chapter has been about the problem of "agency", not in the sense of individual action vis à vis culture and social structure but in its own right. It considered the two most well-developed theories of individual action – learning theory and rational choice theory. Selection processes, learning by reinforcement and punishment in this case, solve the problem of teleology – the logical error which attempts to explain the present by the future. Selection processes like learning theory show that the present is explained as a consequence of the past not by its future – a mode of explanation which has come to be called "teleonomic".

It also described some of the many detailed correspondences that exist between the theory of evolution by natural selection and that of learning by reinforcement and punishment. These included the Hardy–Weinberg equilibrium principle and negative frequency-dependent selection in evolution and two versions of matching in learning; drift and superstitions; adapted or fit and competent; anagenesis versus cladogenesis and shaping versus response differentiation; organismic development-environmental induction and act development-discriminative stimuli; Von Baer's principle and the gradient of reinforcement; adaptive phenotypic plasticity and the cognitive nature of learning; and rule-based genetic and psychological/neurophysiological programmes or algorithms. We also illustrated in Figure 3 how, just as there is a single evolutionary process, there is a single individual learning process which can encompass all of the elementary procedures that yield learning including sensitization, habituation, classical and instrumental/operant conditioning.

It also described the basic assumptions and some of the major predictions of learning's competitor – the "beliefs, preferences and constraints" model of rational choice theory, whether in the narrow (optimization) or broader (including game theory) senses. We described something of the history of its use in the non-economic social sciences. In comparing the theories of learning and rational choice we noted that while neither are teleological in the crude sense, and while just as rational choice cannot explain preferences, learning theory cannot explain why a primary reinforcer is reinforcing. Learning theory, however, does explain the existence of secondary reinforcers, why beliefs come to be as they are, and the emergence of wholly new kinds of behaviours by shaping and response differentiation. Under very restrictive assumptions the predictions of the two theories tend to be similar but because of the aforementioned advantages, and because rational choice assumes a prodigious ability to calculate on the part of individuals, it has been suggested that rationality commonly obtains but as an aggregate phenonemon only, i.e. because of evolution among firms, say, or of learning within individuals. Nevertheless, even in the simplest cases such as a rat learning to turn left in a T-maze, while learning theory in the form of reinforcement is required to explain the direction of change in behaviour – the animal nevertheless began with some preference (e.g. for food) and some belief about where it may be obtained (even if that belief was wrong) as well as some constraints obviously. Apparently then we cannot do completely without either the "forward looking" or the "backward looking" perspectives on individual action. Chapter 8 on "the evolution of complexity" will suggest that this is generally the case in understanding all evolutionary processes including the biological and the sociocultural.

7

Micro and macro II: the problem of subjectivity

7.1 SOCIAL CONSTRUCTIONISM/ CONSTRUCTIVISM

George Ritzer, in the many editions of his sociological theory texts over the years (e.g. 2000), has argued that an objective–subjective continuum (or what many philosophers prefer to term realism versus relativism) is one of the fundamental dimensions along which theories vary. In recent decades, everything from quarks to schizophrenia to gender has been said in a subjectivist or relativist tone to be "socially constructed". Debates surrounding social constructionism or constructivism (used interchangeably) seem destined to mark a watershed in the history of cultural and social theory – not unlike the great functionalism debate of the 1960s. Evidence of this historic importance is the widespread attention they been given by philosophers of science. Consequently, in this chapter I have chosen to pay particular attention to the analysis of four major monographs on the topic written by philosophers (Searle, 1995; Sismondo 1996; Hacking 1999; and Ruse 1999) as well as more recent articles (e.g. Mallon 2009) and they are discussed in the appropriate sections of what follows.

Consider first John Searle's *The Construction of Social Reality* (1995). With respect to the natural sciences and what Searle calls "brute facts" (e.g. the sun is 93 million miles from the earth), he characterizes his position as both ontologically and epistemologically objectivist. He ably defends both realism (the world exists independently of our representations of it) and a correspondence theory of truth (our representations of the world are true if and only if they correspond to the facts in reality). His major project, however, has to do with the social sciences. With respect to what he calls "institutional facts" (those which obtain only when collectively believed that a piece of paper is money or that Bush was president), he characterizes his position as both ontologically and epistemologically subjectivist. Hence the apparent play on word order in the title is apt - not *The Social Construction of Reality* (Berger and Luckmann 1967) but *The Construction of Social Reality* (Searle 1995). Institutional facts take the logical form of "*x* counts as *y* in *c*" as in a piece of paper counts as money in a market transaction. Their uniquely symbolic nature is what makes institutional, unlike brute facts, socially constructed, according to Searle. He associates symbolic representation with consciousness and intentionality and views social structure as composed of hierarchies of institutional facts, which, nevertheless, always rest on a base of brute facts.

Now it is common among semioticians, psychologists and social scientists to distinguish among: (a) a stimulus - which is said to simply cause a behaviour as touching a hot stove causes finger withdrawal; (b) a signal - which is said to indicate probable future events and to acquire its meaning by conditioning processes (as does the bell in Pavlov's salivation experiments or the discriminative stimulus in operant conditioning experiments); (c) a sign - which is said to stand for or represent something else and whose meaning is shared but in which there is an intrinsic connection between the sign and the signified (as in a no smoking notice containing a picture of a cigarette with a line through it, the picture of a cigarette being a sign for a cigarette); and (d) a symbol - which like a sign stands for or represents something else and whose meaning is also shared, but in which there is no intrinsic connection between the symbol and what it signifies (as in a green light meaning go for example). Only human beings it is sometimes claimed interact on the basis of symbolically encoded information in which the connection between symbol and what it signifies is arbitrary. There is some variation from author to author in these definitions. For example Hewitt (2006 and previous editions: chapter 2) omits "sign" in the sense defined above but uses the term "sign" for "signal" in the sense defined above. Despite such minor semantic differences, there is virtually unanimous agreement on symbols - that their meaning is shared and that the connection between sign and signified is not intrinsic but arbitrary. However, the belief that only humans interact on the basis of symbols so defined may be mistaken.

While language is currently believed to be unique to humans, contrary to Searle, symbolically encoded information is not. Symbolic and information-based concepts and language have been in wide use in biology since the genetic code was "cracked" subsequent to Watson and Crick's solution to the problem of the structure of DNA –not only information and the genetic code but also transcription, translation, editing and so on. Since Crick (1968), genomes have been said to be composed of symbolically encoded information because of the arbitrary or historical rather than the physiochemically necessary nature of the connection between a genetic codon (a triplet of nucleotide bases in DNA) and what it stands for or represents (an amino acid in a protein molecule, see Section 5.3). While some would prefer to restrict the concept of symbols applied genetically to the smallest meaningful units (codons) or at least to "genes", others think it appropriate to use it up to the most inclusive functional units. As Maynard Smith (1998:19) put it with respect to the hierarchy of genetic control in development:

> By a 'symbol', semioticians mean a signal whose form is causally unrelated to its meaning. This is clear in the case of words. Thus in English the word 'cow' refers to a particular farm animal. But there is nothing in the sound of 'cow' that makes this meaning necessary – it could equally well mean a mountain or an article of clothing. A few words are not symbolic – 'cuckoo' for instance. But the meaning of most words is conventional. The same is true of genetic signals. The small-eye gene in the mouse means 'make an eye here', but, as far as its form is concerned, it could equally well mean 'make a whisker', or 'don't make a toe.

As a consequence, were non-human organisms interacting socially solely on the basis of genetically inherited information, they would truly and literally be interacting symbolically. Of course the substantive meaning of "no necessary connection" or with a form "causally unrelated to its meaning" in the biological and linguistic cases are different. Biologically there is no physio-chemical necessity and linguistically there is, in addition, no psychological necessity. However, theoretically the point is similar – the association between a symbol and what it symbolizes is a fact of some evolutionary history not a necessity.

Not only is the important difference between humans and other animals not society and not interaction (members of many species live in groups the members of which interact socially), it is not the informational or symbolically encoded basis of the interaction. The important difference in some cases is how that information is acquired – whether via biological inheritance, that is to say genetically, or via sociocultural inheritance, that is to say by social learning (which is how a particular human language is acquired albeit not language in general). In both the genetic and the linguistic cases the meanings of symbols are roughly shared among members of a population – biological and sociocultural populations respectively and in both cases the reason why they are shared is parallel – descent from a common ancestor – biological and sociocultural descent respectively.

Ian Hacking (1999) on the other hand asks us to consider not what is *meant* by social construction but what is the *point*. In surveying many examples of social constructionist research from the literature, he concludes that the point commonly is to raise consciousness. The point is to raise consciousness that something is not biological, nor even inevitable (history is contingent and could have taken a different path), sometimes even that it is bad, and even that it could be otherwise. The point (depending upon the grade of commitment) is often to promote reform, rebellion or even revolution. In short, although he does not use the "p" word, to Hacking, most social constructionist literature in the social sciences is political. This is what Dunn (2006) called "debunking" social constructionism. She distinguishes between two "ontologies" both of which however are about "cultural influence on our understanding". The one is debunking social constructionist accounts which describe only one side of a disputed issue as socially constructed (implicitly implying that the other side is objectively true) and more thorough-going constructionist accounts which describe both sides as socially constructed, the approach which she prefers. An interesting contrast between Searle and Hacking is their emphasis on cooperation and consensus versus conflict and dissensus respectively. To Searle, what is socially constructed is what we agree on; to Hacking it is what we disagree on! Shades of the great functionalism versus conflict theory debates!

There is an actual historical connection between those debates in the 1960s in sociology and anthropology and social constructionism, not because constructionism originally emerged out of one side or the other, but because it originated in reaction against the debate itself. The book that introduced the phrase "social construction", Berger and Luckmann's *The Social Construction of Reality: A Treatise in the Sociology of Knowledge* was published in 1967. It appeared in the context of macrosociological debates over whether societies were (a) akin to organisms or developing organisms, structurally differentiated by a division of labour but functionally integrated and characterized by cooperation and consensus; or whether they were (b) differentiated on the basis of groups historically and structurally unequal in class, power and status with conflicting interests and characterized by conflict and dissensus. Berger and Luckmann's general attitude was "a plague on both your houses". Instead, they focused on the micro over the macro – on how it is interaction among individuals that ultimately creates societies and cultures. It was in the sense of this emphasis on the micro over the macro that they introduced the constructionist metaphor. Along with the earlier work of Irving Goffman, this orientation was quite influential despite the fact that the term "social construction" did not catch on at all. The term was not widely picked up until it took off in the literature in the early 1990s and levelled off in the 2000 to 2005 period (see data in Saioud and Blute 2006). It did so, however, with quite different meanings than Berger and Luckmann's view that, roughly, causality runs from the micro to the macro rather than vice versa.

"Socially constructed" is sometimes used meaning a product of sociocultural rather than biological processes (which is more or less how it is used in Dunn's favoured ontology and by De Block and Du Laing, 2007, in one of the few articles explicitly on its relationship to evolutionary theory). However, if that were all that were commonly implied, then the term "constructed" would hardly be necessary nor would it have provoked such controversy. After all, social scientists have been talking about sociocultural processes at least since the nineteenth century without any apparent need for that term. The same thing would be true, if the only connotation were on an emphasis on the micro over the macro. Symbolic interaction in sociology for example has a history going back to the 1930s (Mead 1934). There is something more or different at issue than these, something which some recent developments in evolutionary theory discussed in the following section as well as some traditional distinctions in psychology discussed in the section after that shed some light on.

7.2   TWO PATHWAYS TO NATURAL SELECTION – PREADAPTATIONS AND NICHE CONSTRUCTION

Natural selection is widely thought of as a sieve, filter or sorting device. For example, Dawkins writes "Each generation is a filter, a sieve: good genes tend to fall though the sieve into the next generation; bad genes tend to end up in bodies that die young or without reproducing" (1995:3). The origin of this metaphor is unknown to me but it is obviously intimately related to the genetical theory of evolution which defines evolution as "a change in gene frequencies in a population" (see any text on population genetics). It has been widely observed that this definition includes genetics and evolution but omits development and ecology. Once these latter are also brought into the picture, it is obvious that the "sieve" metaphor is so simplified that it is positively misleading (what follows in this section is adapted from Blute 2007, 2008a).

Natural selection never acts solely "backwards" as a sieve, filter or sorting device. Instead it always acts sooner or later inductively in a "forward" direction, altering the development of individuals. There are two fundamentally distinguishable pathways. First, an ecological change can induce some individual(s) of a *pre-existing hereditary background* to develop differently whether morphologically, physiologically or behaviorally relative to others. This is possible because phenotypes are plastic (Pigliucci 2001). This in turn can change relative fitnesses, and hence ultimately the frequencies of genetic or other hereditary elements in a population. If a new food source becomes available and is made good use of by some which are hereditably different than others, the former are not just chosen; but changed. They may be induced by their altered nutritional status to grow bigger, live longer or produce more or better offspring for example. Similarly, a new antagonist like a parasite or predator does not bloodlessly choose; it too changes. In this pathway, the sequence is "eco-devo-evo-geno". An ecological change induces a developmental change, which causes an evolutionary change (by changing the relative fitnesses of organisms – it is organisms not genes which survive and/or produce offspring), which causes a genetic change (a change in the frequencies of genetic or other hereditary elements in the population). This pathway (minus the ecological and developmental content) was traditionally called a "preadaptation" in evolutionary theory and thought to be relatively uncommon although more recently under rubrics such as "exaptation" and "co-optation" its probable commoness has been emphasized more.

The second pathway, instead of beginning with an ecological change against a pre-existing hereditary background, begins with a hereditary change such as a new genetic mutation or recombination against *pre-existing ecological background*. A hereditary change leads some individual(s) to develop morphologically, physiologically or behaviourally in such a way relative to others that they perceive, define

or construct a pre-existing feature of the ecological environment differently, changing it, thus changing themselves, thus changing relative
fitnesses, and hence ultimately the frequencies of genetic or other
hereditary elements. If a new hereditary element becomes available
enabling its carriers to consume and make good use of a previously
unutilized resource, the ecological environment is changed. That
change in turn again does not simply sieve, filter or sort but inductively
alters the affected organism(s) improving their nutritional status,
resulting ultimately in a change in gene frequencies. The sequence
here is "geno-devo" then "eco-devo-evo-geno", i.e. a genetic or other
hereditary and developmental change is then followed by the same
eco-devo-evo-geno change sequence as previously. This second pathway implies that niche construction (Blute 1995, Odling-Smee *et al.*
1996, 2003 but in substance see also Hansell 1984, 2000), is not a
once-in-a-while phenomenon. Instead, it is the pathway through
which *all* evolutionary change initiated genetically is achieved.

As a consequence of the need to incorporate ecology and development along with evolution and genetics and because of the existence of
these two distinguishable "inductive" and "constructive" pathways,
Blute (2007, 2008a) suggested the traditional definition of evolution
by natural selection as a change in gene frequencies in a population be
replaced by building on Van Valen's (1973) observation that "evolution
is the control of development by ecology". This yields the following
definition.

> Microevolution by natural selection is any change in the inductive
> control of development (whether morphological, physiological or
> behavioural) by ecology and/or in the construction of the latter by the
> former which alters the relative frequencies of (genetic or other)
> hereditary elements in a population beyond those expected of randomly
> chosen variants.

A third possibility is that both the environmental feature and
the gene pre-existed but what newly emerges is something that *links*
the two. A pre-existing environmental feature might nevertheless
not have activated a pre-existing eating routine until some additional
change on either side resulted in their connection. Rather than "one
gene, one trait" we now understand that there exist whole hierarchies and networks of control of gene activity in development and it
is now believed that more genetic information is dedicated to the
control and linkage of the expression of other genes than is devoted
to structure directly (for an overview see Davidson 2006). One might

reasonably emphasize the "or" over the "and" in the definition above, whether with reference to the simple or the more complex "linkage" case. That is because it seems unlikely that corresponding ecological and genetic changes, whether for good or ill, would take place exactly simultaneously. However, it remains the case that multiple features of the ecological environment are constantly shifting, and in a sexual population, genetic recombination (the source of most hereditary novelty) takes place every generation. Hence, ultimately a particular evolutionary change could be a result not just of one or the other of these exogenously or endogenously initiated pathways or even of the forging of a single linkage between them but of some (any) proportion of all three.

There is nothing in the proposed definition which eliminates the one to many, many to one and requirement for additivity in going from genotypes to phenotypes. It should not escape notice, however, that it does not only encompass both the objective and subjective, but also the ideal and material as discussed in Chapter 5 in the sense that it is a definition that includes both genotypes and phenotypes. While more inclusive than the traditional definition, it is less inclusive than the model of selection processes spelled out in the context of individual learning in Figure 3 (Chapter 6), but is broadly compatible with the latter. For example, biologists have begun to think like psychologists of earlier inductive events "signalling" later conditions with possible mismatches between the two having implications for human health (Bateson *et al.* 2004).

7.3 THE PSYCHOLOGICAL AND THE SOCIOCULTURAL

Consider teaching a rat in a T-maze to turn left as opposed to right when a green light is on. What did it learn? Did it learn the objectively observable behaviour of turning left, to subjectively perceive green, or perhaps simply to associate two pre-existing capabilities? What is the case within individuals is also the case between them. Psychologists have for long tended to avoid talking about "observation" and instead talk about "sensation and perception" (Wolfe 2006; Goldstein 2007). That is because they know that variation in observations can be a function of variation among what is observed, among observers, or even a product of particular classes of observations coming to be linked with particular classes of observers or some (any) proportion of all three. A group of people sitting in a room would probably not have much difficulty in agreeing that some particular object is a table and another one is a chair. Variation in observations in this case, chair versus table, would be wholly attributable to variation in what is observed rather than among observers. On the other hand, a group of people on a bus watching a man in uniform frisking a young black male with the latter's legs spread and his hands against a wall might have a great deal of difficulty in agreeing on what was observed. Some older, white middle-class riders might observe a police officer keeping them safe from crime while others, perhaps younger, black and poorer riders might observe a cop harassing a kid. Variation in observations in this case would be wholly attributable to variation among observers rather than the observed. In most cases, the reality is somewhere between the extremes of objectivity or subjectivity. Still yet, what might have early in life been observations randomly distributed among individuals, could come to be differentially linked in two groups to the internalized consequences of age, race and class. The two examples of chair versus table and "officer doing his duty" versus "police harassment" are not wholly psychological – they are in part sociocultural and economic. A Tuareg who had lived her life in a wholly different cultural and social environment than the people in the room mentioned (for example, in the Sahelian area on the fringe of the Sahara desert with camels, tents and rugs without ever visiting a town or city like Timbuktu or Niamey) might not distinguish among those wooden objects. The observations made by the bus riders might in part be a result of their own biographies but also in part be a result of their inheritance as members of different cultural communities and different social class positions. However much the details may differ, it remains the case that for psychological, sociocultural and/or economic reasons in different cases, variation in what is "observed", can be a result of variation among the observed, the observers, linkage between them or some (any) proportion of each. The social sciences have been down the path of this objectivity–subjectivity issue long before social constructionist language became popular. Are delinquency and madness real or constructed by "labelling" (Becker 1963)? The alternative possibilities exist, no matter how large the scale of the variation/change. In The Affluent Society (1958) and The New Industrial State (1967), John Kenneth Galbraith argued that as the west became more affluent in the post-war period, large firms turned from satisfying needs to creating wants through advertising and marketing. In other words, where once variation in the market environment selected among existing products and firms, with the coming of post-war affluence, variation in products and firms came to construct new market environments.

Hence the issue is not that the subjectivity of observers is not important. It is half the story so to speak. However, according to some, it is not necessary to stop there. One can go on and ask why and how people or animals for that matter have come to perceive, define or construct the world in the particular way that they do. How do some people come to be labelled as delinquent or mad? How did products and firms come to construct new market environments? To their credit, ethnomethodologists (e.g. Garfinkel 1967) have always articulated the question but they also appeared to believe that they had to start from scratch in answering it. In short, there would appear to be every reason to believe that the processes involved in the "evolution" of the programmes governing sensory "inputs" (perceptions, definitions, constructions or whatever) are the same as those involved in governing behavioural "outputs" – the processes of biological evolution, individual learning, and sociocultural evolution. The fact that natural selection, reinforcement, or sociocultural selection does not simply favour alternatives which are well adapted to existing circumstances but also those which construct those circumstances does not therefore vitiate selection principles. In short, selection processes can encompass both ends, both the objective end (e.g. in micro terms "patterns of behavior, action and interaction") and the subjective end (e.g. in micro terms "perceptions, beliefs; the various facets of the social construction of reality") of Ritzer's continuum (2000: appendix, 504).

Mallon (2009) calls these "naturalistic approaches" to social construction. However, if there is no clear demarcation between the "naturalistic" and the objective or the realist then this is what philosophers would call a "metamove" – an objective approach to the subjective or a realist approach to the relativist. Before we accept it uncritically, consider the field in which social constructionism has been most prominent and controversial – in science studies.

7.4   THE SOCIAL CONSTRUCTION OF SCIENCE

The perceptions, definitions and social constructions of the subjects of the social sciences are one thing. Whether they have methodological implications has fuelled myriad debates – introspection versus "interspection" (e.g. Weber's verstehen), emic versus etic perspectives in anthropology, in-depth interviews and ethnographic methods versus polling and statistics in sociology, the debate over appropriating voices in the humanities and so on. Our own as scientists and scholars are quite another matter however.

In science, as elsewhere, subjective perceptions, definitions and constructions evolve – varying and changing historically. An overemphasis on the subjective can be misleading. For example, the neo-Kantian quality of some British and continental European science studies on the "social construction of scientific knowledge" rightly or wrongly conveyed the impression to many that the practitioners of "SSK" believed that the subjectivity of scientists is almost everything and that the objective properties of the natural world contribute little, if anything, to the content of scientific theories. As such it became easy to ridicule: "show me a cultural relativist at thirty thousand feet and I'll show you a hypocrite" (Dawkins, 1995:31–32) and "just try negotiating the AIDS virus into a benign commensal" (Hull, 1994:505). Indeed, Doing (2008) has argued that despite illuminating the all-too-human social life in scientific laboratories, none of the classic laboratory studies (Latour and Woolgar 1979; Knorr Cetina 1981; Collins 1985; Lynch 1985; Pinch 1986) as well as those which have followed have "implicated the contingencies of local laboratory practice in the production of any *specific* enduring technical fact". None has given "an account of a technical fact" as "constructive" rather than "descriptive". He declared that such missing facts are the "dark matter" of science and technology studies. Sismondo, however, (1993, 1996 particularly chapters 4 and 5) found half a dozen uses of the "construction metaphor" in the science studies literature, only one of which conformed to such an untenable subjectivist position. Approaching this literature as one most familiar with traditional "foundational" philosophical analyses of science, he found it refreshing in contributing to an understanding of "science without myth". In the end, he appears to have been drawn to a "non-Panglossian evolutionary epistemology".

Too much of an emphasis on the objective, however, can be equally misleading. In considering whether evolution is a social construction, Ruse (1999) invited us to consider the relative importance of "epistemic" (truth seeking) values such as predictive accuracy, internal coherence, external consistency with the rest of science, unifying power, fertility and simplicity relative to that of other cultural values such as those pertaining to sex, race, religion, and progress in the history of evolutionary biology. He did so by examining the work of a series of prominent evolutionists from the mid-eighteenth century through to the end of the twentieth century. He found an increasing role for epistemic values and a decreasing role for other cultural values through time. Apparently evolutionary biology was indeed once socially constructed but is not, or at least is less so today (albeit metaphors drawn from the broader societal context remain ubiquitous, usefully so according to Ruse.) Scepticism seems likely to greet this discovery of progress. Time (and the work of a lot of historians) has a way of making the importance of the subjectivity of scientists visible in the history of science. By contrast, the absence of such perspective has a way of obscuring the same thing in contemporary science. After all, we are veritably swimming in the sea of the contemporary economic, political and sociocultural context. New science is always technical, difficult and not widely understood until chewed over and made more accessible.

Consider, for example, that one of the two contemporary paragons of epistemic values in evolutionary biology that Ruse discusses at length is the work of Geoffrey Parker. Ruse concentrates on Parker's application of game theory to sexual selection in dung flies and mentions only in passing work for which he is equally well-known – sperm competition. The extremely abbreviated background is that first Darwin discovered sexual selection based on competition for mates as well as natural selection based on competition for survival and fecundity. The concept of sexual selection was more reluctantly received than that of natural selection. Secondly, he discovered (and Huxley named) two types of sexual selection which create two types of characteristics commonly observed in males – intrasexual selection (sometimes called male–male competition) giving rise to weapons like fangs and antlers which are used to fight with and intimidate other males in competition over female mates, and intersexual selection (sometimes called female choice) giving rise to "ornaments" like bright colours and long tails which are attractive to females. The concept of intersexual selection or female choice was more reluctantly received than intrasexual selection or male–male competition. Thirdly, intersexual selection or female choice can be interpreted in two ways. It may be interpreted as the way in which females compete sexually for the best mate (best in a variety of possible senses), i.e. as female choice in a strict sense, sometimes called "active" female choice. Alternatively, it may be interpreted as another way in which males compete sexually (by attracting females), sometimes called "passive" female choice which would more accurately be termed manipulation by males rather than choice by females. The former interpretation was less common than the latter. In short, sexual as opposed to natural selection, female choice in the generic sense as opposed to male – male competition, and active as opposed to passive female choice all received a rough reception and slow acceptance (much but not all of this history is recounted in Cronin 1991).

Parker played the decisive role in the discovery and initial exploration of a whole new level of sexual selection (e.g. 1970). If females mate multiply in a single fertile period, then sexual competition and selection can continue not over mates, but over gametes. Consistent with the history of evolutionary theory, he interpreted this as a form of intra- rather than intersexual competition and selection (as indicated by the very name given it – sperm competition). This was so much the case that, while it may have been less necessary by that time, the editor of a later anthology on the subject felt compelled to follow Parker's review with one "from the female perspective" (Smith 1984:xvii). A later monograph reinterpreting the entire subject was titled *Female Control* and renamed the phenomenon in its subtitle *Cryptic Female Choice* (Eberhard 1996). In a technical monograph, the author devoted most of the first chapter to "previous biases" including eight pages to documenting in detail "the perhaps unconscious assumption that female roles are passive. This is probably part of a general male-centred tradition in biology" (p. 34).

The point here is not who is right and who is wrong. As we saw in chapter 4, both perspectives – male manipulation or female choice, sperm competition or cryptic female choice – may be correct in different or even in the same case. Rather the point is that it is difficult not to see in the history of sexual selection theory in general and as it was continued into the topic of gametes – not simply the influence of useful metaphors drawn from the broader culture – but instead, a long history of the influence of sexist culture, followed more recently by the influence of the cultural "gender wars". Ruse might argue that if the gender wars led eventually to the discovery of "cryptic female choice" so much the better. However, it is obvious that sexist biases in the recognition of sexual selection, of female choice, of active female choice and of cryptic female choice for long impeded rather than facilitated the evolution of knowledge.

Objectivity versus subjectivity or realism versus relativism lie at the heart of constructionist debates. Science in general and sociology in particular is about explaining *differences*, i.e. variation and change. Where there is no difference, there can be no possibility of explaining

why. Why do occupants of some social statuses commit suicide disproportionately relative to those of others? Why did the change to a
capitalist mode of production take place disproportionately in some
religious contexts (if it did) relative to others? However, variation and
change may exist/take place either because the objective behaviour of
individuals or larger social aggregates varies or changes, or because
of how the same objective behaviour is or comes to be perceived,
defined or constructed varies or changes (whether in the eyes of the
actors involved or others), or because different groups of individuals
come to be associated with different views.

The problem enters in when those who emphasize either objectivity or subjectivity try to claim all, instead of only a portion of the
pie. Science simply describes and explains the world as it is. Really?
What of phrenology, Nazi race science, Lysenkoism or cold fusion?
Science simply constructs descriptions and explanations based on, for
example, power politics among scientists jockeying for professional
interest. Really? How did we come to navigate the globe, eliminate
smallpox, or get to the moon? Either camp can also gain enemies
(or friends as the case may be) when the issue in a particular case is
ultimately political as Hacking (1999) suggested. The implicit and
often explicit political message of those who emphasize the objectivity of science is that its institutions need to be defended while that
of those who emphasize the subjectivity of science is that decision
making in science should be opened up to broader participation in the
public interest (more on this in the next section).

Objectivists gain the upper hand by accepting the subjectivists'
point then turning their lens on it. Hence they may admit that the
world is constructed biologically, psychologically and/or socioculturally, but then claim that biology, psychology and/or social science
(implicitly objectively), can describe and explain how it is that something came to be perceived, defined or constructed in a particular way
(Mallon's 2009 "naturalistic approach" to social construction). The age,
race and class status of the observers associated with differences in
what is observed from the bus for example explain those differences.
Much of the social constructionist literature is implicitly objectivist in
this sense. After all, the classical works in science studies were the
result of ethnographic research on scientists working in laboratories.
The converse strategy is a tad more difficult for subjectivists. They
may admit that of course the world is round rather than flat, evolution
happened, and quarks exist, but maintain that such agreement
has been (implicitly subjectively) constructed. There was a time after

all when scientists agreed on no such things and and perhaps they will not again in the future. Few constructionists except perhaps Feyerabend (1975, 1978) go so far as to conclude that scientific knowledge is and should be anarchic, that anything goes. However, because such metamoves can be made in either direction, they appear ultimately unsatisfying as a general principle, even though reasonable people often come to be satisfied about particular cases.

Consider for example the emergence of the link between the HIV virus and AIDS (for a history see Engel 2006). We now know that the HIV virus spread into the human population from some other primate in Africa probably in the late nineteenth century or early twentieth century. However, when a disease syndrome was constructed as "AIDS" (acquired human deficiency syndrome) by the US Center for Disease Control in 1982, the link with the virus was not known. The virus was identified by Luc Montagnier in 1983 and renamed HIV by the International Commission on the Taxonomy of Viruses in 1986, and hence the link between the exogenous in origin HIV virus and the endogenous disease AIDS was made thereafter, "HIV-AIDs" – "human immunodeficiency virus associated acquired immune deficiency syndrome". But evolved linkages may be objective or subjective as well. For some time thereafter, most famously Thabo Mbeki, then President of South Africa, and his health minister, but also many others and not only in South Africa, argued that the linkage was a social construction. Eventually in 2002 they were over-ruled by the South African Cabinet and appropriate retroviral treatment was introduced in South Africa in 2003. Few any longer deny the objectivity of the link. The outcome of the story would be quite different if we were to delve into some other scientific controversies – such as cold fusion for example.

Stephen Cole, one of the first generation offspring of the founder of the modern sociology of science, Robert Merton, in *Making Science: Between Nature and Society* (1992) adopted a sensibly moderate position between extreme objective or realist and subjective or relativist positions in science studies. He calls himself a "realist-constructivist" and explains why he:

> believes that science is socially constructed both in the laboratory and in the wider community, but that this construction is influenced or constrained to a greater or lesser extent by input from the empirical world. Instead of saying that nature has no influence on the cognitive content of science, a realist-constructivist says that nature has *some*

influence and that the relative importance of this influence as compared
with social processes is a variable which must be empirically studied.
I do not believe that evidence from the external world *determines* the
content of science, but I also reject the position that it has no influence.

More recently, Harry Collins (2009) one of the early relativists in
science studies who had introduced scepticism about the role of
"decisive experiments" in science (1985), writing in an essay in
Nature, also advocated a more balanced view.

7.5 THE POLITICS OF SCIENCE STUDIES

Because much of the constructionist literature is political as Hacking
showed, we should consider some of the political conclusions that
have been drawn explicitly or implied by the science studies literature. According to the late John Ziman (2000), an English physicist
turned philosopher and sociologist of science and always an acute
observer of the scientific scene, there has been a gradual transformation in science. This transformation has been from the "academic
science" which prevailed from roughly the mid-nineteenth century
when the term "scientist" was invented, to "post-academic science"
which began to develop in the 1960s and has accelerated since
then. The culture of academic science was characterized by Merton's
(1942) norms of communalism, universalism, disinterestedness,
originality and scepticism which Ziman dubs "CUDOS". It was
financed by generous donors and governments; carried out by individuals free to set their own research agendas and to publish; built
on previous work by others; motivated by recognition in their scientific communities; and was largely carried out in universities and
institutionally organized there, as well as more broadly in scientific
and scholarly societies in disciplines. In return grateful scientists,
through peer review systems, provided their societies and governments in particular with credible, reliable, reproducible knowledge
and the latter drew on them for expertise and advice.

By contrast in "post-academic science", the norms of "PLACE"
have come to replace those of "CUDOS" – science is propriety, local,
authoritarian, commissioned and expert. Knowledge may not be made
public, work is done on local technical problems, governed by a managerial hierarchy, commissioned to solve specific problems and the
scientist is valued as a technical expert rather than for creativity.
Post-academic science may be carried out in a privatized setting or in

a university, but in either case, it is industrialized with a high division of labour. According to Ziman, there is a reason why sceptical views of science which scorned traditional claims of its "disinterestedness" became popular in this latter period – because science has indeed become less so. Hull (2001b) writing in *Nature*, called Ziman's picture of contemporary science "detailed", "realistic" and "well-rounded". While Ziman's description of the changes which science has undergone is offered in the flat, neutral tone of seemingly objective social science, reaction to such changes has been anything but.

Objectivists or realists tend to be conservative in supporting traditional "academic" science. They are alarmed for example by the extent to which entrepreneurial scientists and science administrators have allied their science with special interests – once commonly said to be the defence industry but more recently the pharmaceutical industry has been signalled out. Conservative supporters of traditional science are shocked by practices such as the contracting out of clinical trials to "don't ask, don't tell" firms who provide "science for sale"; the publishing of papers whose results they like and the shelving of those they do not; the use of public relations firms to "manufacture uncertainty" about harmful products; and undisclosed financial interests and "ghost" (read fake) authorship of papers by academic scientists and other similar scandals. Rather than broadening participation, they think that controversial issues in science should be settled in the peer-reviewed literature and sometimes when necessary, by consensus conferences of the credentialled – the assessment reports of the Intergovernmental Panel on Climate Change are good examples. They believe that the future of traditional academic science dedicated to the acquisition, preservation and free dissemination of knowledge is by no means guaranteed. Harking back to our first two chapters, particularly if science is an historical inheritance (a clade or groups of clades) rather than a phenomenon which, in nomothetic style, recurs whenever conditions are right (a guild), the organizations and institutions which originally embodied it such as universities, scientific societies, disciplines, journals and peer review need to be defended. That is because with clades, extinction is forever. As well as traditional bodies such as Royal Societies and National Academies of Sciences, there are even popular organizations like the Committee for the Scientific Investigation of the Paranormal, Free Thought Associations and popular publications like *The Skeptical Inquirer* more or less dedicated to the defence of science. The view of one of the most-informed observers of the science scene (Greenberg

2007) is that while the dangers are very real, "campus capitalism" is commonly not as profitable as some believe and that scientific institutions have not been entirely helpless in defending themselves (journals coming to require pre-registration of clinical trials so that the publishing of results cannot be selective for example).

Subjectivists or relativists tend to be radical in their attacks on science. They emphasize that generations of philosophers have failed to successfully demarcate science from non-science – science from pseudoscience or even from religion say. Since what is considered scientific truth is largely a matter of power politics within the laboratories and the institutions of science which, like the rest of society, are stratified on class, gender and racial lines for example, they would like to see a weakening of scientific credentialism and closed decision making in science opened up to broader public participation in the interests of making science come to better serve the public interest. Fuller (2006: chapter 6) for example suggests a model of "citizen science" in which consensus conferences based on the model of citizen-jurors be used to decide science policy. They can be coy at times about whether they are talking about the fields and topics of research that should be supported and what scientific discoveries should be allowed to be put into practice (cloning, genetically modified food, etc.) or whether they are talking about decisions about actual scientific questions. There is no doubt that at least some mean to include the latter. Fuller (2006: chapter 5) would like to see creation science taught by creation scientists alongside evolution science taught by evolution scientists in universities for example. Interestingly, he is willing to defend the professional expertise of historians, philosophers and sociologists of science (Fuller 2008).

7.6   SUMMARY AND CONCLUSIONS

Debates over social "constructionism" or "constructivism" (used interchangeably) in the social sciences have attracted the same kind of attention from philosophers of science in recent decades as did the functionalism debate of the 1960s. Searle (1995) proposed essentially that physical facts are not socially constructed but that sociocultural ones are – the kind on which it takes agreement to make them facts like that certain kinds of paper are money. He associates this with the uniquely human capacity for the use of symbols. However, not only human language, but also information encoded genetically is widely understood by biologists to be symbolic because of the historically evolved rather than physio-chemically necessary link between a genetic codon and the amino acid it codes for. If some animals were interacting solely on the basis of genetically encoded information then, that interaction would be symbolic. To Hacking (1999), something which is only socially constructed is not something we agree upon, but the opposite, something we disagree upon. Debunking social constructionist arguments are political; they are about something that someone thinks did not have to develop the way it did and should be changed. To some, social constructionism is only about sociocultural rather than biological processes, or within the former, only about an emphasis on micro over macro processes. However, given their long history, neither of these required "constructed" language nor would they likely have stirred such controversy.

Science is about explaining differences – change or variation. In contemporary evolutionary theory change can be introduced either because some (external) environmental change (by inducing developmental, fitnesss, and hence gene frequency change) selects among pre-existing heritable variation, or because some (internal) change like a genetic mutation or recombination leads some individual(s) to "construct" the pre-existing environment differently thereby altering it (which in turn induces developmental, fitness, and hence gene frequency change), or even because something external and internal become linked. This suggests a new definition that "microevolution by natural selection is any change in the inductive control of development (whether morphological, physiological or behavioural) by ecology and/or in the construction of the latter by the former which alters the relative frequencies of (genetic or other) hereditary elements in a population beyond those expected of randomly chosen variants." Similarly, in learning, a rat in a T-maze rewarded for turning left when a green light is on may have learned to objectively turn left, to subjectively perceive green or to associate the two. Similar possibilities are available in the social sciences. Variation in "observations" among individuals can be a result of variation among the observed, among the observers or even because particular classes of observations come to be linked with particular classes of observers or some (any) proportion of all three. Metamoves are common in debates over objectivity versus subjectivity or realism versus relativism – we can explain subjective differences (implicitly objectively, Mallon's 2009 "naturalistic approach" to constructionism) or we can explain objective differences (implicitly subjectively).

Because both metamoves are possible, both are ultimately unsatisfy-ing as a general rule, which does not mean that reasonable people do not often come to be satisfied about particular cases such as the objectivity of the link between HIV and AIDS for example.

Nowhere has the debate over constructionism been more fierce than in the "science wars". That case seems to fit Hacking's analysis – the argument is ultimately political. Conservative defenders of tradi-tional science fear the extent to which society in the form of the defence and pharmaceutical industries, for example, have intruded into academic science. Radicals think society in the form of public participation in decision making in the public interest has not intruded nearly far enough.

The answer to the question of whether everything, something or nothing is socially constructed is first that everything is not "socially" so – some things are surely biologically or psychologically constructed. Latour and Woolgar later altered the subtitle of their 1979 classic *Laboratory Life* from *The Social Construction of Scientific Facts* to *The Construction of Scientific Facts*. Secondly, whether talking biology, psychology or social science – everything is not constructed, some things are constructed, and most things are partially constructed. Thirdly, ideally, neither side should try to claim everything. Is the link between the AIDS virus and HIV really subjective? Is the biological literature on human gender differences really objective? If science needs to be defended against anything, it needs to be defended against both extreme objectivism which would have us believe that subjective perceptions, definitions and constructions can be ignored and extreme subjectivism which would have us believe that they are everything. Stephen Cole (1992) and more recently Harry Collins (2009) adopted a sensibly moderate position. As to whether science needs to be defended from society, or whether society needs to be defended from science, perhaps some of both is called for.

8

Micro and macro III: the evolution of complexity and the problem of social structure

8.1 INTRODUCTION

The concept of "progress" hovers about all questions pertaining to complexity but since progress is inherently a normative or evaluative concept it cannot be answered scientifically. The scientific problem of complexity can be divided into three broad categories – ecological, individual and social. The problem of ecological complexity is what determines the number of kinds of things (species), how new kinds arise (the process of speciation), and whether there is an overall trend toward increased complexity as they do. These are discussed in Section 8.2 below.

The problem of individual complexity is the problem of, apart from the issue of more kinds and trends – when new, more complex kinds of individuals *do* arise, how they do so. Biologically, how did life get from prokaryotic cells to the eukaryotic to multicellular individuals to eusocial colonies? How did human societies get from lineages to clans to tribes to nations to empires? How did weapons get all the way from spears to Intercontinental Ballistic Missiles and joint stock companies all the way from the East India Company to behemoths like Exxon Mobil, General Electric and Wal-Mart? This is discussed in Section 8.3 below.

The problem of social complexity is a kind of half-way house between these, the problem of how interacting social groups coalesce but also disaggregate so that while selection on a higher level may be involved, nevertheless they do not form new permanent aggregate individuals. The most notable example is the formation of sexually reproducing families from asexual individuals discussed in chapter 4. We described there the role of byproduct mutualism in the form of the advantages to individuals of ecological specialization but also the advantages of bet-hedging to aggregates, i.e. to sexually reproducing families, populations and species. However, we noted that it could probably equally well be described in terms of Okasha's (2006) "contextual analysis" model of group selection in which causal cross-level byproducts may obtain in both upward and downward directions between "particles" and "collectives" and that new aggregates can be formed through conflict as well.

Section 8.4 is about how everything that evolves also develops and that includes the sociocultural including social roles, statuses or identities for example. In this section we ask what is required for new aggregate levels of selection to evolve in this dynamic, developmental sense – what is required for the evolution of replication, e.g. for culture to become recursive. Lastly in Section 8.5 we explain why we think it is appropriate to talk about the sociocultural rather than the cultural and the social separately and delve into the polythetic meaning of "social structure".

8.2   PROGRESS AND ECOLOGICAL COMPLEXITY

The concept of progress is inherently normative or evaluative. If you value more, bigger or more complex, then if these kinds of things evolve, you would call that progress. If you value fewer, smaller or simpler, then if these kinds of things evolve, you would call that progress. Because such normative or evaluative questions cannot be answered by science, the majority of biologists deny believing in progress in evolution although Ruse (1997) has shown that many evolutionists, including Darwin himself, have implicitly believed in progress in evolution more than they like to admit. Social scientists too have generally rejected the concept of progress in cultural evolution in that sense as well – that later emerging languages are somehow "better" than earlier ones for example.

The species concept does not make a lot of sense in the absence of eukaryotic sex and because of the complexity of syngamy and meiosis, it is generally believed that the latter arose de novo only once. (It has re-emerged in a very, very few species which have lost it, assumed to be because the capacity to produce males remained latent – Domes et al. 2007.) Although most species that ever existed are extinct, and history has been punctuated by mass extinction events, there has obviously been a long-term increase in ecological complexity in the sense of the number of species (about 1.5 million are currently named). Cultural complexity increases in this sense as well. If hammers or computers were invented once or a few times, there is currently a profusion of diversity of kinds of them (on the former see Basalla 1988:4–5 after Blanford). These can be readily accounted for, however, by a "left wall" – if you start with one or a few and change is going to take place, there is only one direction to go and that is up. A more interesting question is about the rate of speciation – do species numbers tend to increase linearly for example? The best evidence currently suggests that the rate of biological speciation within clades is density-dependent, i.e. the rate of speciation slows down as geographical and ecological space becomes limited (Phillimore and Price 2008). The same thing is likely to be true culturally. For example, when a new drug is introduced (and its patent runs out), it is commonly followed by a bunch of "me-too" entries into the market (Spector 2005), but the room for such should become exhausted. By analogy with within-species competitive strategies, this might seem to suggest that the members of later emerging species within a clade should be efficient rather than productive consumers and quality rather than quantity producers, i.e. that they should be larger, and if larger is correlated with more complex (which some microbiologists would dispute) it would seem they should be more complex as well. We will return to that question after we consider the process of speciation itself.

"Good" species tend to differ on where they live, the ecological niche that they occupy and how they look and behave, particularly including the fact that their members do not interbreed with those of others. With speciation, a wedge has been driven into a population or species separating it and the major theories of speciation differ on the nature of that wedge. The wedge may be physical ("allopatric" speciation), ecological ("sympatric" speciation) or social (by means of sexual selection and/or hybridization). The "allopatric" (different homelands) theory is said to be spatial or geographic. However, when examined it normally means that the wedge is some kind of a physical barrier. The barrier may be newly risen (the movement of continents, the rising of mountains, rivers taking new courses or even a new rut in a muddy road for snails) or it may exist because some individuals accidentally cross a pre-existing, usually impenetrable barrier such as the water between the mainland and an island. In either case, the physical isolation precedes changes that subsequently take place through somewhat different selection pressures or genetic drift on either side of the barrier which are such that if the populations again come into contact, they are unable to successfully interbreed with each other. Allopatric speciation is common culturally as well. Speakers of the same language are able to exchange communications with each other in more or less the same way that members of the same biological species are able to exchange genes but speakers of different languages and members of different biological species are not. Cavalli-Sforza (1997, 2000) found positive correlations between small genetic polymorphisms and the languages spoken by peoples around the world, not because the genetic differences caused the linguistic differences or vice versa, but because historically the same barriers that tend to isolate biological species also tend to isolate human languages and cultures. Not surprisingly, mountainous regions of the world such as Papua New Guinea tend to be rich both in species diversity and in human linguistic and cultural diversity.

The "sympatric" (same homeland) or niche theory maintains that natural selection stemming from ecology, i.e. interaction with other species, can itself drive wedges into populations and species, even if inhabiting the same geographical area. This was assumed by Darwin (Reznick and Ricklefs 2009) who, after all, called his book *The Origin of Species*. However, under the influence of Ernst Mayr (1942) it was long thought impossible by biologists because they assumed that interbreeding would break down such differences. While some debate continues (Bolnick and Fitzpatrick 2007; Fitzpatrick, Fordyce and Gavrilets 2008), a growing number of cases are recognized (Schilthuizen 2001). Insect species of a common ancestral origin may disperse to and live on different species of plants even though the plants are interspersed in the same geographical area for example. Social scientists recognize the plausibility of sympatric speciation. Dialects of a language are incipient languages and are normally allopatric, but what are variously called "registers" or "sociolects" of languages and "subcultures" of a culture are normally sympatric. Lawyers and businessmen in their suits, students in their jeans, hip hop boys in their baggy pants and so on mingle in the subways and main commercial areas of my city and not only are they distinctive, but they rarely interact and even if they tried to do so, their chatter would be virtually incomprehensible to each other. The key to sympatric speciation in both biological and sociocultural cases is that "speciation is a process and not an event" (Grant and Grant 2007), and as divergence proceeds for other reasons, reduced interaction proceeds right along with it. The former can even favour the latter, a process biologists call "reinforcement".

The wedge can be physical or ecological but it can also be social. In the last chapter, we considered the case of ecological differences between proto-genders and genders followed by sexual selection. But what if the ecological differences first took hold within a gender? Then sexual selection in a variety of ways including passive or active mate choice by the other gender could drive a population apart. While there remains debate over how common it is likely to be (Ritchie 2007), it is now considered likely that sexual selection is responsible for the rampant sympatric speciation which has taken place among cichlid fish in East African lakes (Knight and Turner 2004). Finally, speciation by social means can be brought about by two wedges and a clamp rather than a single wedge. Occasionally hybridization between species can give rise to a new species which is reproductively isolated from both its ancestral species, particularly in plants. Many social scientists have observed that hybridization is more common culturally than it is biologically – think of the way the many genera of world music now hybridize into new forms. It should be noted, however, that cultural intermingling, like the biological, is far from being unconstrained. As I once noted, "One may get an idea for a handle shape for a pot from a basket, but when was the last time characteristics among tables and curtains or staplers and rugs say were recombined? Obviously in the world of culture, differences among populations can and do eventually become deep enough that the gap is rarely if ever bridged" (1979:55).

Today in the attempt to answer the question of which process is involved, attention tends to be focused on the very earliest stages. It seems unlikely that there will ever be a single universal theory of "the origin of species" but that does not mean that there may not someday be a small number of universally specified theories – under these conditions this kind of speciation takes place, under those conditions that kind and so on.

Now let us return to the question of trends. Obviously there has been an increase in the *maximal* size and complexity of cellular organisms over the course of history as there has been of cultural entities such as weapons, firms, etc., but again that is simply explicable because of a "left wall". If life or some technology or social form began with small and simple examples and if change was to take place, there was nowhere to go but up. Daniel McShea and colleagues in a series of studies and publications have been asking whether the members of new species, when they arise, tend to be more complex than those of previously existing ones. Their most interesting question is what happens *on average* – does evolution tend to drive complexity upward? Major transitions in complexity in evolution (viewed as nestedness – prokaryotes within eukaryotes, cells within the multicelled, individuals within colonies) have been too infrequent for statistical analysis. Instead McShea and colleagues have studied "minor transitions", the degree of "individuation" of higher level entities, specifically the degree of connectedness among parts, their differentiation, and the existence of intermediate-level parts such as tissues and organs. While the methods and analysis are too complex to summarize here – the bottom line was that they most recently found "no pervasive tendency for organisms to become more complex hierarchically. Rather our findings are consistent with the notion that the trend is the result of diffusion away from a lower boundary (Marcot and McShea 2007:199)." While it has not been so carefully investigated, it is likely the same thing is true culturally. For example, just because anthropologically we can identify lineages, clans, tribes, nations and empires – that does not mean most lineages expanded into clans, most clans into tribes, most tribes into nations and most nations into empires. Undoubtedly most lineages that ever existed became extinct, as did most clans, and tribes. A few remained as "living fossils" and a few went on to increased complexity.

However, how can we reconcile the evidence from McShea and colleagues against cross-lineage trends towards increasing complexity with the argument that, because speciation is density-dependent, later-evolved species should tend to be bigger and probably more complex than earlier ones. The answer would seem to be that new species tend *not* to be competing with each other. If they did, perfectly, i.e. in all respects, according to the competitive exclusion principle, they could not coexist. There is a limit to the number of "me-too" drugs a market can support, for example. Instead, new species along with becoming reproductively isolated, normally diverge from each other in the way they act and the niche they occupy and that is the point of all three kinds of speciation. The allopatric do not compete for spatial reasons, the sympatric do not compete for ecological reasons, and the sexually selected do not compete for ecological reasons within one gender and social reasons within the other.

There are numerous other questions about ecological complexity that the life and social sciences have in common and that could be pursued further. For example, species diversity is greatest in the tropics and declines toward the poles as does linguistic and cultural diversity. Both biological species and human languages and cultures are currently in a period of mass extinction (Sutherland 2003) that is anthropogenic, i.e. brought about by humans (Nettle and Romaine 2000). Biologists tend towards the view that ecological communities which are more diverse are more stable, but some evidence has recently been brought forward that initial evenness in relative abundance is what makes for stability (Wittebolle *et al.* 2009). Social scientists too would like to believe that multiculturalism and equality are good things, but whether the former tends to stabilize or destabilize nation states remains in doubt. Both life and social scientists are interested in movements into new geographical areas – biological "invasions" and socio-cultural "migrations" – and the conditions that make for success or the lack thereof of such migrants.

8.3   INDIVIDUAL COMPLEXITY

Just because there may be no overall trend in evolution towards greater complexity does not mean that this "vertical dimension" of evolution as Arthur (2006) called it is not interesting. Maynard Smith and Szathmary (1995:3) claimed: "The most that we can say is that some lineages have become more complex in the course of time", but that did not lead to disinterest on their part, quite to the contrary. Several potentially important ideas about the nature and evolution of individual complexity have emerged in recent years. By far the most notable on the nature of complexity has come from molecular developmental genetics (including the comparative) which has revealed that the control of the expression of genetic information is organized in vast hierarchies and networks, some of which extend surprisingly far across the tree of life and the history of which can be studied by comparative methods. The writings of Sean Carroll (Carroll, Grenier and Wetherbee 2000; Carroll 2005, 2006) have made this story quite well known.

But other significant ideas about the nature of complexity have emerged as well. One is *modularity*. This stems back to Ohno's (1970) discovery that new genetic sequences often emerge by duplication and divergence of old ones. But this can take place at any level from the genetic to body segments for example and is a large part of what makes the evolution of individual complexity possible. This is because as complexity increases, interaction effects between parts are liable to become so complex that "evolvability" is impaired. The "many to one" (genetic epistasis) and "one to many" (genetic pleiotropy) phenomena can make organisms literally "unevolvable" – any novelty inevitably has too many negative side effects. When duplicated, however, duplicates can be maintained as they go on to evolve to serve new functions, and importantly such units can become semi-autonomous, a phenomenon sometimes called "near decomposability". The idea is that interaction effects between modules can be suppressed making possible the evolution of individual complexity (see Callebaut and Rasskin-Gutman 2005 for a variety of discussions). Formal organizations do this all the time, also at a variety of levels, multiplying departments, divisions and so on often creating semi-autonomous "profit centres".

The single most important idea about the evolution of complexity introduced in the past decade or so was that of Maynard Smith and Szathmary (1995:3) who viewed the existence of complex individuals as a result of "major transitions in evolution". Such major transitions have in common that "entities capable of independent replication became able to do so only as part of a larger whole" (prokaryote cells incorporated into eukaryotes, protists in multicellular individuals, and solitary individuals in colonies with non-reproductive casts, for example). The components of new levels of organization and selection in their view could emerge by duplication, symbiosis or epigenesis. They thought that the advantage to individuals, similarity, a tendency to irreversibility, and many to suppress the selfish could commonly explain such transitions but often the advantages of a division of labour – "the efficiency of specialized organs" as well as the emergence of "new materials and mechanisms of heredity" (p. 12) were required as well.

Conceived in this way, the problem of individual complexity is more or less the same as the problem of the evolution of cooperation discussed in Chapter 4. How is cooperation among individual replicators in new aggregates achieved and enforced? As such, the potential answers are more or less the same as those discussed there pertaining to sex (by cooperation and/or conflict) except that in most of the products of major transitions, unlike the case of sex, there is no temporary aggregation and disaggregation, but the evolution of new permanent aggregates. The advantages of specialization among different types of "parts" for example is part of the same explanation offered for the evolution of social complexity in the form of sexual families in Chapter 4.

Social science deals with a world with more than one and perhaps multiple levels of selection as well. Individual behaviours, norms and values can obviously be transmitted, vary, be selected and evolve. So too with these aggregated in social roles, statuses or identities. Inheritance among the latter may not be neatly asexual (one to one) or sexual (two to one) but like does tend to produce like – it is members of a descent group in traditional societies who socialize the young into membership in the group, it is a group of doctors who as members of the faculty of a medical school socialize new doctors, church members who socialize new members and so on. Multi-status or multi-role aggregates are more difficult to judge. Are they members of reproducing and evolving populations or are they ecological aggregates? It varies. Medieval monasteries and Hutterite colonies did or do reproduce, vary and evolve. Both had or have multiple different kinds of roles or statuses within them and communities of monks and colonies of Hutterites fission when the local resource base is outgrown – both being careful to include the variety of necessary occupations in each offspring colony. Phylogenies of medieval monasteries are available in histories (Knowles 1963). The difference between them is that Hutterite colonies generally socialize their own biological offspring into their roles while medieval monasteries generally recruited the biological offspring of others to socialize. Some other entities with multiple statuses and roles may appear superficially like they reproduce but do not. Franchise operations and pyramid sales schemes are examples. When you acquire a McDonald's franchise, you acquire the right to sell McDonald's hamburgers but not the right to sell the right to sell McDonald's hamburgers. Formal organizations differ from institutions in that while both include multiple different kinds of statuses with different roles, the former normally have an origin at a particular point of time and some kind of charter with a stated mission and mode of governance. Although there is a considerable literature on the evolution of formal organizations such as for-profit enterprises (e.g. McKelvey 1982; Singh 1990; Baum and Singh 1994; Aldrich 1999, 2006; Durand 2006), organizations are not normally in the business of making more little organizations. The origin of replication was discussed in Blute (2006b) and this case in particular was discussed in Blute (2007). The conclusions drawn were that while firms exist in populations, are born, develop competitively – competing to "grow" their business in a particular niche, display many organismic like properties such as some adaptive flexibility and die, selection among them is normally viability-based only. However, selection based on viability or competitive development only can lead to a surprising amount of evolution, probably because as time goes on, the range of variation present in new foundings increases. Even more significantly, recently some have begun to divide, split or spin off divisions which are truly independent (unlike branch plants, franchises, etc.) so that in this case we are literally seeing a new level of selection emerge in front of our eyes so to speak. It was predicted that increasingly the success of firms in the future will be judged by their relative success at this kind of proliferation.

Maynard Smith and Szathmary transformed the old issue of the units and levels of selection and the possibility of group selection into an historic one as Okasha (2006) noted – major transitions. However, it is also a dynamic question in a second, developmental sense. Because everything that evolves also develops – new aggregate entities must undergo a development process in a competitive context, one which replicates, a topic to which we now turn.

8.4 COMPETITIVE DEVELOPMENT OF NEW LEVELS OF REPLICATION: RECURSIVE CULTURE

Replication is normally achieved biologically by bequeathing offspring genetic instructions and the material means with which to implement them in the expected environment. Commonly it is achieved socio-culturally by transmitting linguistic instructions and the material means with which to implement them in the expected environment. But what of the content of the programme? For simplicity's sake we consider an example which assumes that consumption (including digestion) and offspring production (including quality) by collective individuals in a population of collectives are density-dependent and that the niche is full. If the initial state is many, small individuals, selection favours consumption. Some grow as others die off constructing a population with a few, large individuals. This favours switching to offspring production. Numbers grow as sizes shrink, constructing a population with many, small individuals favouring a switch to consumption again. Collectives which are plastic in the form of responsive to reliable cues in this way would be favoured over those which only consume, or only produce offspring or do both but under the opposite conditions. Such collectives would be favoured in the sense that they would be recursive, to use Griesemer's (2000) term, i.e. replicate their life cycle. This particular case is a semelparous life cycle (offspring production takes place once at the end of the life cycle). Viewed from the perspective of the particles in such collectives and cross-level byproducts upwards, as a minimum, collectives must include two kinds of particles that interact in such a way that each constructs the ecological (here demographic) conditions which favour (and presumably, in a developmental context, induce) the other, resulting in replication of the collective (see Figure 4). This then is the criteria for the *competitive development of a new level of replication, variation, selection and evolution.* It incorporates the same general kind of relationship that we saw previously in Chapter 4 between a minimum of two complementary strategies which must obtain between males and females spatially

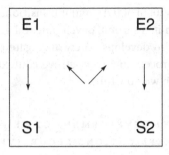

* Vertical arrows are induction, slanting arrows are construction. An ecological condition E1 favours and induces the development of strategy S1 which constructs ecological condition E2 which favours and induces the development of strategy S2 which constructs ecological condition E1 again and thus results by competitive development, of a new level of replication, selection and evolution. See text for full explanation.

Figure 4. Competitive development of new levels of replication.

to replicate sexual families but is here required temporally, i.e. developmentally, to replicate collective individuals asexually. In such a case, both particles (in the long run) and collectives benefit. There need not be a complete absence of conflict among particles, between somatic and reproductive tissue for example, only that conflict must be controlled such that the cost of any conflict which does take place be less than the advantages of specialization so that particles not in collectives are not advantaged relative to those which are. Of course, with different assumptions and starting points a variety of other kinds of life cycles in addition to the semelparous are possible including the iteroparous and what might be described as a "reverse iteroparous" – when offspring are born larger than their parents and reproduce before they grow. Similarly, the primordial complementary alternatives in multicellular organisms need not be consumption and production although that may be the most fundamental even in uni-cells (Turke 2008) – they might be feeding and motility (as with normal versus flagellated choanocyte cells of sponges) or feeding and digestion (as with the ectoderm and endoderm of diploblasts) for example.

While this kind of an explanation can explain how it is possible in the more inclusive sense for new levels of development to be repeated, i.e. to replicate, it does not fully explain how their development actually proceeds in a mechanistic sense. How does this mutual ecological induction and construction involving complementary strategies take place during development? If Maynard Smith and Szathmary are right that *all* new levels of complexity are composed of evolving entities at a lower level (prokaryotes in eukaryotes, cells in multicellular individuals, individuals in eusocial colonies, etc.) then the development of aggregate individuals *could not proceed by any processes other than normal evolutionary ones at that lower level*. Evolution on the lower level becomes the mechanism of development on the higher (Blute 1977; Buss 1987). In multicellular development for example, the cells and cell groups that express the strategies which have such complementary consequences for the collective of which they are a part must be being induced and constructing by means of heritable change (whether genetic, e.g. copy number changes or epigenetic), by being acted on and acting by means of selection pressures exerted on one another, and by engaging in motility in such a way that the collective life cycle as described is brought about. We might thus describe development as an encapsulated evolution, which is not to imply that the ecological environment including the demographic is out of the loop. On this view, whoever said that evolution never repeats itself could not be more wrong. It repeats itself virtually every time a collective individual in a lineage replicates.

We saw in Section 2.3 that social learning cannot be reduced to psychological processes and in Section 5.6 not obviously to social psychological processes either. For example, to perpetrate a collective social identity like membership in a church rather than simply a single behaviour, norm or value, ego must normally provide alter linguistic instructions (and resources) on how to consume, to model herself on ego, and to produce, to instruct her recruit(s) to do likewise with respect to her, as well as about the appropriate demographic conditions under which to do each. From the perspective of the population of these sociocultural identities, when there are a lot of recruits but none of them know very much about what it means to be a church member, they need to concentrate on consuming, i.e. modelling themselves on existing members. However, when there are only a few recruits but ones who are well socialized into what it means to be a church member, they need to concentrate on producing, i.e. recruiting new members by teaching. Only by such flexible means will membership be truly socioculturally recursive, i.e. replicate socioculturally. From a mechanistic perspective however, this consuming requires "learning to learn" and producing requires "learning to teach", both of which must be taking place at a lower, i.e. psychological level within the same individuals (which as we have seen is also a selection process) and by way of social psychological interactions among them. This is unlike the case in Section 5.6 in which different individuals learn to teach and to learn. So the psychological and social psychological are part of what is going on but by themselves are not sufficient – the sociocultural ecology including the demographic matters too.

Debates over levels of selection – whether such concepts are illegitimate, just different ways of looking at and talking about the same thing, or necessary for real understanding – date at least from the 1960s and show no evidence of having abated (e.g. Sarkar 2008). Where some see different levels, others see simply different scales. I expect more rather than less of this as evolutionary and developmental biology are brought more into interaction with each other as is taking place in "evo-devo". Among other reasons, that is because viewing selection as taking place at higher levels or more inclusive scales has necessary implications for understanding development at the lower or less inclusive.

8.5 THE SOCIOCULTURAL: CULTURE, SOCIAL RELATIONSHIPS AND SOCIAL STRUCTURE

Cultural anthropologists normally historically spoke of cultural evolution while sociologists spoke of social evolution. The two are combined in the title of this book (as they often are elsewhere) because the distinction is ultimately a false one. Once anthropologists got out of the third world and sociologists got out of the first world, they have had trouble distinguishing their subject matters. An almost political treaty was signed between the heads of these two tribes when Alfred Kroeber and Talcott Parsons (then presidents of the American Anthropological and the American Sociological Associations respectively) co-authored a joint article (1958) declaring that "culture" is the subject matter of anthropology and "social relations" that of sociology. Although some recent evolutionary social theorists (such as Runciman 2001, 2002; Klüver 2002, 2003; Rousseau 2006) have thought that we require a theory of the evolution of social relationships distinct from that of culture, I disagree. The units of culture are or should always be specified in relational terms. A normative characteristic of the role of a dentist is to minimize pain in his *patients*, of a scientist to consider disinterestedly explanations for phenomena offered by his *peers*, of a mother to behave benevolently towards her *children* and so on. It is not the case, however, that the relational quality of cultural norms means that some second or additional subject matter "relationships" exist as a subject matter of the social sciences. It would seem, in other words, that a social relationship is fully described when the cultural content of the roles involved in it are fully specified, including the "towards whom". Reciprocally, a full description of a social link requires a specification of the cultural content of the nodes so to speak. This should not be taken to imply a rejection of the topic of social evolution as discussed in Chapter 4 which included the content of both roles and a description of their interaction, just of the existence of "links" independent of the properties of nodes. Nor for that matter should it be taken to imply that patterns of relationships such as the block models of social network analysis, for example, are not interesting or worthy of study.

No term is used more ubiquitously in sociology than "social structure" and no term is less well or less commonly defined. Sometimes the term social structure is used simply to reflect our perception that the distribution of culture (and the distribution of resources including wealth and income, power and prestige among its units) is not random. It is patterned. It exists and changes in patterns of associations which are potentially explicable by social science. In this broadest sense possible, social structure is just a way of talking about everything that social scientists want to study and explain. In this sense, social *structure is everything*. More generally, however, it is much easier to say what structure is *not* than what it is. Structure is not culture, not change, not variation, not history, not function, not agency, not construction, and given that we are talking about social science, not biology.

As discussed in the opening paragraph of this section, in the eyes of some, social structure is not culture. It is not culture in a second sense as well as discussed in Section 5.7. Talk of social structure tends to emphasize the material basis of existence over ideas. In both of these senses, *structure is not culture.*

It was common some decades ago to provide two second year courses in sociology programmes – one on "social structure" and one on "social change". The synchronic and diachronic were neatly divided up by courses! Courses on social structure were, by and large, descriptive. They described how behaviours and beliefs and the norms and values governing them tend to be clustered in social roles, statuses or identities which are commonly transmitted as a package (along with associated resources such as wealth and income, power and status). They described how these in turn are aggregated in organizations and institutions. Change then was studied separately, commonly from either from a functionalist or from some variant of a neo-Marxist perspective. In this sense, social *structure is not change*. A related sense of structure was that employed by Claude Lévi-Strauss and his followers which had its historical roots in de Saussure's (1916, 1966) distinction between "langue" (language) and "parole" (speech). On that view, behind the diversity of appearances of speech or kinship systems or myths for example, lie a smaller number of archetypes, forms or essences such as de Saussure's language and Lévi-Strauss's (1969) classification of systems of marriage alliances. This kind of essentialism is recognizable to biologists in debates over the nature of species for example. However misguided "essentialism" may or may not be (it has after all been reinvented by cladism after having been banished from biology), it commonly does result in a great deal of very useful work of description and classification which can then be interpreted in an evolutionary framework as can Lévi-Strauss's work on kinship for example. When European theorists talk of being "poststructuralists", they commonly mean having abandoned this kind of structuralism as well as Marxism. In any event, in this sense *structure is not variation*.

In Chapter 3 on necessity, we distinguished between the weight of history and, after Durkheim, the "social forces" that push and pull people around and sometimes this emphasis on contemporary causes, essentially on the selective forces that permit or prevent the maintenance of something or act for or against its spread, is called "structural". In this sense, *structure is not history*. Structure is often contrasted with function. Robert Merton, probably the greatest essayist in the history of sociology, once disagreed with one of his mentors at Harvard, Talcott Parsons, over the latter's functional theory of society. He did so in a way that is largely familiar to evolutionists. Merton (1968) argued that functionalism needed to be modified into "structuralfunctionalism". He criticized the postulate of functional unity (something may not be functional for all parts of society), the postulate of universal functionalism (there needs to be a calculation of the net balance of functions and dysfunctions) and the postulate of functional indispensability (alternative structures may be able to perform the same function). Because functionalism tends to emphasize cooperation and consensus in society, talk of structure tends to emphasize conflict. Anyway, *structure is not function*.

Historically both sociology and anthropology, both functionalist and Marxist, tended to emphasize how societies and cultures affect individuals rather than how individuals affect societies and cultures.

Table 5. *Social structure as a cluster concept*

Structure versus	Structure emphasizes	Over	Chapters
Culture	the material	the ideal	5(7)
	relationships or links	properties of nodes	8(3)
Change	stability	change	
Variation	"essence" similarity	"accidents" variation	8,9
History	necessity	history	2,3
Function	conflict	cooperation	4
Agency	reinforcement	rational choice	6
Construction	objective	subjective	7
Biology	sociocultural	biological	9

Talk of structure tends to implicitly emphasize reinforcement over rational choice and ultimately the sociocultural over either of these. In both of these senses (the second of which has been left to the final chapter to discuss) *structure is not agency.* Talk of structure also tends to emphasize how the world is objectively, rather than how it is perceived, defined or constructed subjectively. In this sense, *structure is not construction.* Finally, because we are talking about social scientists' use of "structure", *structure is not biology.*

For all of these reasons, I would suggest that the concept of social structure is a polythetic or cluster concept rather than one for which necessary and sufficient conditions for its application can be stated, even theoretically, let alone empirically. The concept of social structure tends to emphasize one side of most of the major issues in cultural and social theory discussed in this book. It tends to emphasize the material over the ideal; stability over change; necessity or social forces over history; conflict over cooperation; reinforcement over reason and the sociocultural over individual agency; the objective over the subjective, and given that we are talking social science, the sociocultural over the biological (see Table 5).

8.6   SUMMARY AND CONCLUSION

Progress is an inherently evaluative concept, questions about which cannot be answered by science. Complexity can be ecological, individual or social (an example of the latter, the evolution of sexual families, populations and species was discussed in Chapter 4). Although punctuated by mass extinctions, there has been a long-term increase in

ecological complexity (number of species biologically and socioculturally) but the rate of speciation within a clade is density-dependent,
i.e. declines with crowding. This might lead us to expect the later
evolved to be larger and more complex. New species can emerge for
physical, ecological or social reasons and while the maximum size and
complexity have increased because of a left wall, the evidence is that
there is no average trend towards increased complexity in evolution.
The density-dependent expectation that there should be and the fact
that there is not can be reconciled if new species generally tend not to
compete with the old whether for physical, ecological or social reasons.

Because there is no average trend towards increased complexity,
it is still of great interest when it does occur. Ideas introduced in recent
years about increased individual complexity in addition to molecular
developmental genetics include those of modularity and of major
transitions in evolution. In major transitions entities capable of independent replication become able to do so only as part of a larger whole
(such as prokaryote cells incorporated into eukaryotes, protists into
multicellular animals, and solitary individuals into colonies with non-
reproductive casts for example). This historical approach to increased
complexity needs to be supplemented by a developmental approach,
i.e. the new aggregate entities need to undergo a life cycle which
replicates, i.e. is recursive. This can be achieved by the same general
approach taken to the evolution of social complexity in Chapter 4, by
the existence of a minimum of two complementary strategies now
within aggregate individuals, each of which constructs ecological
such as demographic conditions which favour (and in the developmental context, induce) the other. A biological and sociocultural example of
the evolution on a new level of a semelparous life cycle is provided. It
was suggested that viewing selection as taking place at higher levels or
more inclusive scales has necessary implications for understanding the
mechanism of development at the lower or less inclusive.

A firm distinction between the cultural and the social and hence
between cultural and social evolution is not justified, which is why this
book is titled Darwinian *sociocultural* evolution. Social structure is rarely
defined but appears to be a polythetic or cluster concept which emphasizes one side of the many issues in cultural and social theory discussed
throughout this book.

9

Evolutionism and the future of the social sciences

9.1 INTRODUCTION

Those who agree with the centrality of evolution to the social sciences do so with two quite different meanings in mind. One is the gene-based biological. The other, as has been emphasized here and by others (e.g. Van Parijs 1981; Hull 1988; Luhmann 1995; Hodgson 1999; Ziman 2000; Wheeler, Ziman and Boder 2002; Lenski 2005) is the social learning-based sociocultural. Beyond either lies the incredibly complex tangle of how the two affect and interact with one another as well as with the learning/choices of individuals. While some modest progress has been made – there remain more questions than answers.

9.2 THE BIOLOGICAL

The genus Homo goes back about two and a half million years. For long it was believed that the last remaining species in this genus apart from us, *Homo sapiens*, was *Homo neanderthalensis* which died out in Europe about 25 000 years ago. What would seem to be the greatest scientific discovery of the early years of this new century was that of the remains of some members of a third *Homo* species, *Homo floresiensis*, small and very small-brained but tool-using, which may have survived up to 12 000 years ago on the island of Flores in Indonesia (Brown *et al.* 2004). While there was some disagreement about its species status as opposed to humans afflicted with a disease, the weight of evidence currently seems to be in favour of the former (Falk *et al.* 2005; Jungers *et al.* 2009). To put this remarkable discovery in perspective, we are talking about a time about as close to when humans began to build cities in Mesopotamia (3000 to 4000 years ago) as we are to when the Neanderthals died out! Hints of subsequent incipient speciation among humans exist. For example the

"Pygmies" of central Africa appear to share a unique common ancestry about 2800 years ago – more or less at the time of the expansion of non-pygmy agriculture, which tended to confine the former to the forests, perhaps favouring a short stature and creating further differentiation among isolated populations of them. Subsequently, to varying degrees in different groups, there has been (largely male) gene flow from non-pygmy to pygmy populations (Verdu *et al.* 2009). The bottom line, however, is that all humans today are members of one biological species, potentially capable of interbreeding with one another. It seems inconceivable that unless a nuclear, climactic, impact, disease or other kind of catastrophe isolates remaining small populations of humans for a long period of time, or unless we some day colonize other planets, that evolution in the sense of cladogenesis is in the cards in the future for humans.

Evolution in the sense of gene-frequency changes and differences, however, are another matter. Durham (1991) first drew widespread attention to the probability that herding cultures selected for genes enabling the digestion of the milk sugar lactose into adulthood in certain human populations, which was confirmed (e.g. by Burger *et al.* 2007). More recently, Perry *et al.* (2007) showed the same thing for farming and genes for the digestion of starch (both also reviewed in Patin and Quintana-Murci 2008). In the former case the change was in the level of expression and in the latter by copy number expansion, but in both cases a variety of mutations arose which were rapidly convergently selected in several places in the world and which remain different in different geographical areas with different dietary histories. It is virtually certain that evolution in this sense of gene frequency changes did not cease when people were all hunter-gatherers and indeed presumably continues to take place in the human species to this day. In fact, the argument has been made that human adaptive evolution has been *accelerating* in the past 40 000 years (Cochran and Harpending 2009). There is genomic evidence for this (Hawks *et al.* 2007) but some caution is in order with respect to existing methods of identifying genomic evidence for adaptive evolution (Hurst 2009). However, it makes sense that the spread of humans out of Africa into a great diversity of ecological environments, disease epidemics in denser populations, the demands of acquiring multiple social identities and negotiating multiple different kinds of social relationships in those dense populations, and last but not least, coevolution with culturally transmitted "memes" should have accelerated the rate of adaptive evolution (Section 9.4 below).

9.3   THE SOCIOCULTURAL

There is a long history in science fiction of imagining a world in which products of human technology acquire the capacity to reproduce, escape, and become a threat to ourselves. The tradition began with Samuel Butler's *erewhon* (an anagram for nowhere), first published in 1872, through Asimov's (1950) *I Robot* (in which the robots were programmed with three rules to keep them under control, however), to Michael Crichton's (2002) fantasy *Prey* about nanobots run amok. That there may actually be a danger in the near future has been taken more seriously by some scientists, engineers and assorted techies since Bill Joy (2000), a co-founder of Sun Microsystems, published an essay titled "Why the future doesn't need us" in *Wired* magazine early in the new century. In this confessional essay on his once perhaps misguided faith in science and technology, Joy warns of the possibility of "knowledge-enabled mass destruction" reducing the biosphere, including ourselves, to "a gray goo". The danger lies in replicating and evolving "assemblers", products of new technologies whether in "robotics, genetic engineering or nanotechnology".

In the light of this, it is interesting to reflect on the fact that, according to most social scientists, that world is already here – it arrived long ago with culture. Not necessarily that culture is always biologically maladaptive (see Section 9.4 below), but that culture long ago came to swamp the biological in the determination of human behaviour, mind and social organization. It is difficult for biologists to understand that anthropologists, sociologists and political scientists are not really interested in "people" per se – those skin-bound bundles of flesh and blood. They view the world with a metaphysic of sociocultural not biological realism. Hence they are interested in "slices" of people – in them as occupants of particular social roles, statuses or possessing particular social identities. They are interested in doctors or entrepreneurs or plural wives or voters or suicide bombers or evangelical Christians or criminals. They are also interested in how behaviours are clustered under the control of norms and values governing such social identities; in how these roles or statuses are organized in formal organizations (whether public or private, non-profit or for-profit); in how individual roles or statuses and formal organizations are organized in institutions such as kinship, religious, economic, and political institutions; in how roles or statuses, organizations and institutions are organized in a state; and how states relate to each other. They are particularly interested in how resources – wealth, income, power and status are distributed at all of these levels, the cooperative and exploitative relationship that exists among them, and how and why they all remain stable or change.

The great majority of social scientists therefore are not likely to become intensely interested in biological evolution or even gene-culture coevolution. Despite this, I still believe that before another century is out the majority of social scientists will come to echo Dobzhansky who once famously declared that "nothing in Biology makes sense except in the light of evolution" (1973) by coming to see that nothing in social science makes sense except in the light of evolution either. For these traditional and exclusively *social* scientists, evolutionary theory provides a *model* for understanding which is attractive because of its inclusiveness. As I hope to have demonstrated in this book, it does not force us to choose between the ideal and the material; change and stability; history and necessity; cooperation and conflict; reason and reinforcement; the subjective and the objective, or ultimately between the biological and the sociocultural. Logically and empirically, it link the poles of such dichotomies. As claimed in the preface, it is a general theory or metanarrative which shows how what we all have to say makes sense. However, for those life and social scientists who *are* interested in the relationship and interaction between genes and culture there will be fertile ground to plough.

9.4   INTERACTION: GENE–CULTURE COEVOLUTION

It is often thought that, in the interaction between genes and culture, genes have to be given pride of place because, after all, they came first. Presumably individual learning/decision making followed next (particularly in non-social species) and social learning came last (in social species). However, our understanding of the relationship among these need not be governed by their presumed order of appearance. While contrary to some memeticists I do not think it is generally useful to think of culture as a virus, the analogy is useful in this particular respect. Mainly because viruses do not possess ribosomes (protein factories), they are obligate intra-cellular parasites. But that does not mean that once having emerged, in some cases almost certainly as escaped parts of cells and perhaps in other cases as reduced cells themselves, that they are then not evolving in their own right. And so it is with culture. Because culture is a "second inheritance system" in Boyd and Richerson's expressive phrase, sociocultural evolution interacts with gene-based biological evolution on an equal footing so to speak.

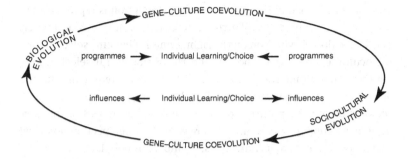

* Outer loop – biological evolution on left and sociocultural evolution on the right. But they also coevolve in interaction with each other in the outer loop – genetic elements select among sociocultural ones along the top, sociocultural items select among genetic ones along the bottom (i.e. gene-culture coevolution obtains). Middle section – upper line, both genes on left and culture and social organization on the right (and their coevolutionary interactions) can programme individuals in the middle to learn/choose in particular ways. Middle section – lower line, there are many issues involved in the nature of the influence individual learning/choice exercises on the genetic on the left and on the cultural on the right (and on their coevolutionary interactions) – in particular whether those influences can or should be conceived of in informational or material terms. See text for full explanation.

Figure 5. Interactions among biological and sociocultural evolution and individual learning/choice.

There exists a model for the relationship between genes and culture in the biological model of coevolution between interactng species. Lumsden and Wilson (1981) first used the term "coevolution" to refer to the interactions between genes and culture. Swanson (1983) talked of the interaction between "socioenes" and "biogenes" in the human dual inheritance system. When Boyd and Richerson (1985) first enquired about what difference the existence of individual and social learning make to the biological evolutionary process, they devoted at least a few pages (pp. 194–197) to the possibility of antagonistic coevolution between genes and culture. Durham (1991) labelled the case of genes selecting among cultural alternatives as "genetic mediation" and of cultural alternatives selecting among genes, "cultural mediation". He illustrated cultural mediation with malaria resistance (with sickle cell anaemia as its side effect) selected for by slash and burn agriculture, and with the ability to absorb lactose into adulthood, selected for by dairying cultures. Blute (2006a) employed a phenotype-based game theory logic to the relationship suggesting that genes and culture may evolve mutualistically (a plus plus relationship, most likely with vertical transmission), antagonistically (a plus minus relationship, most likely with horizontal transmission), or competitively (a minus minus, relationship most likely with oblique transmission). Gene–culture coevolution is represented in Figure 5 by the outer oval of arrows. Biological evolution is on the left and sociocultural evolution on the right. Genetic elements select among sociocultural ones along the top and the sociocultural select among genetic ones along the bottom. The additional complexity in the gene–culture coevolution case is first that both genes on the left and socioculture on the right programme the learning/choices of the *same* individuals (upper middle line) whereas in the interspecific biological case, these are different individuals, members of populations of different species. So life scientists have to ask what difference does the existence of cultural transmission and evolution make to the biological evolutionary process and social scientists have to ask what difference does the existence of biological transmission and evolution make to the sociocultural evolutionary process.

9.5 INDIVIDUALS IN THE MIDDLE

A second complexity, one not restricted to gene–culture coevolution in particular but shared by both is the nature of the influences exercised by the learning/choices of individuals on both evolutionary processes and on their coevolutionary interactions (Figure 5, bottom middle line). A key issue here is ultimately whether those influences should be conceived of in informational or material terms and whether that matters, a matter which requires some explanation.

While they tend understandably to close ranks when confronted with enemies like intelligent design creationism, biological evolutionary theory is in somewhat of a turmoil currently. The concept of the gene is, as one colleague who will remain anonymous put it to me, "frankly, a mess". As discussed in Section 5.2, that problem is unlikely to be solved until the problem of genetic recombination is solved. But its existence in no small measure is responsible for a rebellion in some quarters against the use of all informational concepts, which at the other extreme, are at the heart of and seem to be an indispensable part of molecular genetics, including of development. A second source of turmoil is discontent with the inclusion of only evolution and genetics and neglect of development and ecology in the traditional population genetics understanding and definition of evolution as "a change in gene frequencies in a population". I offered the alternative definition quoted in Section 7.2:

> Microevolution by natural selection is any change in the inductive control of development (whether morphological, physiological or behavioural) by ecology and/or in the construction of the latter by the

former which alters the relative frequencies of (genetic or other) hereditary elements in a population beyond those expected of randomly chosen variants

as a contribution to better integration of evolution, heredity, development and ecology.

Despite the fact that it clearly recognizes that evolutionary innovation can be a matter of environments changing the expression and hence the fitness of old genes as well as new genes acting constructively on old environments, the more radical critics are unlikely to be satisfied by any such moderate adjustment. Older concepts such as the Baldwin effect, Waddington's genetic assimilation and developmental constraints and new ones like the ontogeny of information and developmental systems (Oyama 2000), genetic accommodation (West-Eberhard 2003), facilitated variation (Kirschner and Gerhart 2005) and compositional evolution (Watson 2008) are being discussed. Many of them include echoes of the outsiders' criticism – that gradualist, random gene-centred evolution is unable, despite Dawkins' (1996) claims, to "climb mount improbable". At the risk of simplifying, they tend to think that the variation side of the "variation and selection" theme (with the roots of variation in development) requires more elaboration if the introduction of novelty and the evolution of complexity are to come to be truly understood. Whether any of these are tinged with Lamarckianism I leave it to others to judge. It is important to understand that Lamarckian *inheritance* in the sense of the inheritance of acquired characteristics is no problem. It is ubiquitous. Whenever a cell doubles in size and divides, however, the new material is distributed among the offspring, *fifty percent* of what is inherited by them collectively was acquired rather than inherited by the parent. Lamarckian *evolution* in the sense of the preferential inheritance of acquired adaptations over acquired maladaptions however is a problem. Frankly, it would be miraculous.

I have consistently defended the view here (Chapters 1 and 2) and elsewhere that sociocultural evolution is Darwinian rather than Lamarckian because there is no evidence that sociocultural innovations, as a statistical body, are oriented in the direction required for them to successfully spread. Now we get to the point of the problem of the individual in the middle. There is no doubt that biological evolution can programme individuals and individuals can affect biological evolution. But is there any sense in which those latter influences can be understood in information as opposed to purely material terms? The same issue arises in sociocultural evolution. There is no doubt that cultural evolution can programme individuals and that individuals can affect sociocultural evolution. But is there any sense in which those latter influences can be understood in informational as opposed to purely material terms?

At least some think there is. In sociology, Dennis Wrong (1961) once famously complained about the "overly socialized concept of man" in sociology. It is fairly common today for arguments like that of Swidler (1986) to be heard in the social sciences according to which culture is a "tool kit" from which people "construct their strategies of action" particularly in "unsettled lives". I hazard that most of those who study either individual learning or rational choice or both discussed in Chapter 6 would agree with the Swidlers of the world. This is the real problem of "agency", not of reason versus reinforcement as discussed there, but of "agency" versus culture and social structure. It would not be fair to conclude this book without informing the reader that the question of the role of the individual in both biological and sociocultural evolution is currently under discussion.

9.6 LANGUAGE A TARGET?

The series of wrenching dominant paradigm shifts which linguistics has undergone in its history are ample evidence of the fact that understanding the relations among genomes, individuals and culture is not going to be easy. In the first phase, historical linguistics dominated. It viewed language as a cultural phenomenon in which languages within a family (and perhaps beyond) descended with modification in a tree-like fashion from a common ancestral language by a process of descent with modification. For a time the psychologically oriented looked like it might vie for the mantle of successor, the type-specimen being Skinner's *Verbal Behavior* (1957). However, it was quickly displaced by an essentialist form of structuralism which had its historical roots in de Saussure's (1916, 1966) distinction between "langue" (language) and "parole" (speech). On that view, behind the diversity of speech, (sentences in a language, languages in the human species) there exists a common form, essence or archetype – language. The type-specimen here became Chomsky's (1957) transformational-generative grammar which came to be interpreted biologically by Pinker (1994) and others as representing the human language instinct. For a time a gene-culture coevolutionary hybrid originating with Deacon (1997) which tried to make peace between the biological and sociocultural with slogans such as "brains evolved for language and language evolved for brains" began to emerge and may still eventually hold sway. During much of this time, the psychological in the form of the study of language development in individuals has lurked in shadows, not unlike the way early mammals lurked in the shadows in the age of reptiles. Current understanding of evolutionary processes suggests that individual development will be a crucial link in whatever version of "language evolution" comes to dominate. In the meanwhile, Everett (2005) began another revolution with the argument that the last bastion of universalism in language, "recursiveness" (Hauser *et al.* 2002), has fallen, being absent in Pirahã, and has led to a new challenge from the sociocultural (Evans and Levinson 2009). I predict that Evans and Levinson's massive article will become a citation classic rivalling Chomsky's (1959) skewering of Skinner. For convenience I have described the evolution of dominant paradigms in linguistics as a stage-like sequence, but as any evolutionist would expect – culturally oriented historical linguistics did not disappear with the coming of the psychologically oriented, the psychologically oriented with the coming of the biologically oriented structural, the biologically oriented with the coming of the coevolutionary, and the coevolutionary will be unlikely to disappear with the revival of the culturally oriented again. Instead they all persisted in some form and even underwent their own internal diversifications but the point is that, in successive periods, newly emerged groups did become more ecologically dominant.

Nevertheless, it is in the field of language evolution that a synthesis may most likely be first obtained. This is so for a number of reasons. First, in substantive terms, language is the single most important component of culture and is the one in which much of the rest is embodied. Secondly, there are vast amounts of data available to be collected or analysed secondarily. Thirdly, social organization matters and the subject of "evolutionary linguistics" in the form of the interdisciplinary EVOLANG conferences have been held every two years since 1966 and are heading into their eighth in 2010 at Utrecht (for their flavour see the Christiansen and Kirby anthology 2003; the list of proceedings is available under evolutionary linguistics on Wikipedia). "The Evolution of Language" would appear to be deliberately ambiguous – origin or maintenance? biological or sociocultural? collectively or individually? In addition, it includes people who have a sophisticated understanding and even do research on individual learning as well. Practitioners of many disciplines sharing their findings and theoretical insights with each other and doing so about a particular subject like language is probably what is needed to move the synthesis forward. One such attempt at synthesis is Tomasello (2008) whose theme is a dual biological and sociocultural evolutionary one, but there will be many more efforts before this task is completed.

9.7   SUMMARY AND CONCLUSIONS

Biological evolution continues in the human species although speciation seems unlikely. Most social scientists whose view it is that "culture has people rather than people having culture" as a colleague put it, are unlikely to become interested in biological evolution or gene–culture coevolution. However, evolutionary theory can, and will, it is predicted, become more and more useful for understanding their subject matter in its own terms. Beyond the two processes considered separately, there exists the complexity of their coevolutionary interactions and the relationship between both and the individual. The process of coevolution between species provides a model for thinking about the former but both have the additional complexity of dealing with not just how genes and culture programme individuals, but in what sense the latter affect the former. In the social sciences, this is the real problem of the role of human agency in a culturally programmed and socially structured world. Linguistics, which has seen a wrenching series of dominant paradigm shifts in its history, may be one of the most fertile areas in the future for studying evolution in the multiple senses of "the evolution of language" because a vibrant tradition of interdisciplinary interaction has emerged.

References

Abell, Peter. 2000. Sociological theory and rational choice theory. Pp. 223–244 in Bryan S. Turner, Ed., *The Blackwell Companion to Social Theory*, 2nd edn. Oxford: Blackwell Publishers.

Adams, Cecil. 2000. Was standard railroad gauge (4'81/2") determined by Roman chariot ruts? *The Straight Dope.* http://www.straightdope.com/columns/000218.html

Adams, Julia, Elisabeth S. Clemens, and Ann Shola Orloff, Eds., 2005. *Remaking Modernity: Politics, History, and Sociology.* Durham, N. C.: Duke University Press.

Agrawal, Aneil F. 2006. Evolution of sex: Why do organisms shuffle their genotypes. *Current Biology* **16**:R696–R704.

Albert, Victor A., Ed., 2005. *Parsing Phylogeny and Genomics.* Oxford: Oxford University Press.

Aldrich, Howard. 1999, 2006. *Organizations Evolving.* London: Sage Publications Ltd.

Allen, Nicholas J., Hilary Callan, Robin Dunbar and Wendy James, Eds., 2008. *Early Human Kinship: From Sex to Social Reproduction.* Oxford: Blackwell Publishing.

Andersson, Malte. 1994. *Sexual Selection.* Princeton: Princeton University Press.

Archer, Margaret. 1995. *Realist Social Theory: The Morphogenetic Approach.* Cambridge: Cambridge University Press.

Ardrey, Robert. 1966. *The Territorial Imperative: A Personal Inquiry into the Animal Origins of Property and Nations.* New York: Atheneum.

Arnqvist, Göran and Locke Rowe. 2005. *Sexual Conflict.* Princeton: Princeton University Press.

Arthur, Wallace. 2006. *Creatures of Accident: The Rise of the Animal Kingdom.* New York: Hill and Wang.

Asch, Solomon E. 1951. Effects of group pressure upon the modification and distortion of judgment. Pp. 177–190 in Harold S. Guetzkow, Ed., *Groups, Leadership and Men: Research in Human Relations.* Pittsburgh: Carnegie Press.

Asimov, Isaac. 1950. *I Robot.* New York: Bantam Books.

Aunger, Robert, Ed., 2000. *Darwinizing Culture: The Status of Memetics as a Science.* Oxford: Oxford University Press.

Aunger, Robert. 2002. *The Electric Meme.* New York: The Free Press.

Avise, John C. 2006. *Evolutionary Pathways in Nature: A Phylogenetic Approach.* Cambridge: Cambridge University Press.

Axelrod, Robert. 1984. *The Evolution of Cooperation.* New York: Basic Books.

Axelrod, Robert. 1997. *The Complexity of Cooperation: Agent-Based Models of Competition and Collaboration.* Princeton: Princeton University Press.

Axelrod, Robert. 2006. *The Evolution of Cooperation: Revised Edition*. New York: Basic Books.

Bailey, Ronald. 2001. Post-scarcity prophet: Economist Paul Romer on growth, technological change, and an unlimited future. *Reason Magazine*, December.

Ball, Philip. 2002. Paper trail reveals references go unread by citing authors. *Nature* **420**:594.

Ball, Philip. 2008. A longer paper gathers more citations. *Nature* **455**:274–275.

Bandura, Albert, Ed., 1971, 2007. *Psychological Modeling: Conflicting Theories*. New Brunswick, N. J.: Aldine Transaction.

Bandura, Albert. 1977. *Social Learning Theory*. Englewood Cliffs, N. J.: Prentice-Hall.

Bandura, Albert. 1986. *Social Foundations of Thought and Action: A Social Cognitive Theory*. Englewood Cliffs, N. J.: Prentice-Hall.

Barbrook, Adrian C., Christopher J. Howe, Norman Blake and Peter Robinson. 1998. The phylogeny of The Canterbury Tales. *Nature* **394**:839.

Barnard, Christopher J. 1984a. *Producers and Scroungers: Strategies of Exploitation and Parasitism*. London: Croom Helm Ltd.

Barnard, Christopher J. 1984b. The evolution of food-scrounging strategies within and between species. In Christopher J. Barnard, Ed., *Producers and Scroungers: Strategies of Exploitation and Parasitism*. London: Croom Helm Ltd.

Barnard, Christopher J. and Richard M. Sibly. 1981. Producers and scroungers: A general model and its application to captive flocks of a house sparrow. *Animal Behaviour* **29**:543–550.

Barry, Andrew. 2003. Review of W. D. Hamilton's Narrow Roads of Gene Land, Volume II, The Evolution of Sex. *London Review of Books* **25**: Feb. 6.

Bartholomew, Robert E. and Erich Goode. 2000. Mass delusions and hysterias: Highlights from the past millennium. *Skeptical Inquirer* **24**:20–28.

Basalla, George. 1988. *The Evolution of Technology*. Cambridge: Cambridge University Press.

Basalla, George. 2006. *Civilized Life in the Universe: Scientists on Intelligent Extraterrestrials*. New York: Oxford University Press.

Bateman, A. J. 1948. Intra-sexual selection in Drosophilia. *Heredity* **2**:349–368.

Bateson, Patrick, David Barker, Timothy Clutton-Brock *et al.* 2004. Developmental plasticity and human health. *Nature* **430**:419–421.

Baum, Joel A. C. and Jitendra V. Singh, Eds., 1994. *Evolutionary Dynamics of Organizations*. New York: Oxford University Press.

Becker, Carl L. 1932. *The Heavenly City of the Eighteenth Century Philosophers*. New York: Yale University Press.

Becker, Gary S. 1993. Nobel lecture: The economic way of looking at behavior. *Journal of Political Economy* **101**:385–409.

Becker, Howard. 1963. *Outsiders: Studies in the Sociology of Deviance*. New York: The Free Press.

Bell, Graham. 1978. The evolution of anisogamy. *Journal of Theoretical Biology* **73**:247–270.

Bell, Graham. 1982. *The Masterpiece of Nature: The Evolution and Genetics of Sexuality*. Berkeley: University of California Press.

Bennett, Charles H., Ming Li, and Bin Ma. 2003. Chain letters and evolutionary histories. *Scientific American* June:76–81.

Benzer, Seymour. 1957. The elementary units of heredity. Pp. 70–93 in W. D. McElroy and B. Glass, Eds., *A Symposium on The Chemical Basis of Heredity*. Baltimore: Johns Hopkins Press.

Berger, Peter L. and Thomas Luckmann. 1967. *The Social Construction of Reality: A Treatise in the Sociology of Knowledge*. Garden City, N.Y.: Doubleday.

Berrill, Neil J. 1976. *Development*. New York: McGraw-Hill Book Company.

Bertalanffy, Ludwig von. 1968. *General Systems Theory: Foundations Development and Applications*. New York: G. Braziller.

Best, Michael L. 1997. Models for interacting populations of memes: competition and niche behavior. *Journal of Memetics* **1**.

Best, Michael L. and Richard Pocklington. 1999. Meaning as use: Transmission fidelity and evolution in NetNews. *Journal of Theoretical Biology* **196**:389–395.

Beurton, Peter. J., Ralph Falk and Hans-Jörg Rheinberger. 2000. *The Concept of the Gene in Development and Evolution: Historical and Epistemological Perspectives*. Cambridge: Cambridge University Press.

Bichakjian, Bernard H. 2002. *Language in a Darwinian Perspective*. Frankfurt: Peter Lang Publishing.

Bielby, J., G. M. Mace, O. R. P. Bininda-Emonds *et al.* 2007. The fast-slow continuum in Mammalian life history: An empirical reevaluation. *The American Naturalist* **169**:748–757.

Birdsell, J. A. and C. Wills. 2003. The evolutionary origin and maintenance of sexual recombination: A review of contemporary models. *Evolutionary Biology* **33**:27–38.

Blackmore, Susan. 1999. *The Meme Machine*. Oxford: Oxford University Press.

Blaffer Hardy, Sarah. 2009. *Mothers and Others: The Evolutionary Origin of Mutual Understanding*. Belknap/Harvard University Press.

Blau, Peter M. 1964. *Exchange and Power in Social Life*. New York: John Wiley.

Blute, Marion. 1977. *Darwinian Analogues and the Naturalistic Explanation of Purposivism in Biology, Psychology and the Sociocultural Sciences*. PhD thesis, University of Toronto.

Blute, Marion. 1979. Sociocultural evolutionism: An untried theory. *Behavioral Science* **24**:46–59.

Blute, Marion. 1982. Evolutionary and ecological processes in marketing: The product life cycle. Pp. 71–74 in Ronald F. Bush and Shelby D. Hunt, Eds., *Marketing Theory: Philosophy of Science Perspectives*. Chicago: American Marketing Association.

Blute, Marion. 1995. The New Dawkins: Review of Richard Dawkins, *River Out of Eden: A Darwinian View of Life*. *Trends in Ecology and Evolution* **10**:504–505.

Blute, Marion. 1997. History versus science: The evolutionary solution. *The Canadian Journal of Sociology* **22**:345–364.

Blute, Marion. 2001a. Social learning by observation is analogue, instruction is digital. *Behavioral and Brain Sciences* **24**:327.

Blute, Marion. 2001b. A single-process learning theory. *Behavioral and Brain Sciences* **24**:529–531.

Blute, Marion. 2003. The evolutionary ecology of science. *Journal of Memetics* **7**(1) http://jom-emit.cfpm.org/2003/vol7/blute_m.html

Blute, Marion. 2005. If the genome isn't a God-like ghost in the machine, then what is it? *Biology and Philosophy* **20**:401–407.

Blute, Marion. 2006a. Gene-culture coevolutionary games. *Social Forces* **85**:151–166.

Blute, Marion. 2006b. Origins and the eco-evo-devo problem. *Biological Theory*, **1**:116–118.

Blute, Marion. 2007. The evolution of replication. *Biological Theory* **2**:10–22.

Blute, Marion. 2008a. Is it time for an updated 'Eco-Evo-Devo' definition of evolution by natural selection? *Spontaneous Generations: A Journal for the History and Philosophy of Science* **I**(2),1–5. http://jps.library.utoronto.ca/index.php/SpontaneousGenerations

Blute, Marion. 2008b. Cultural ecology. Pp. 1059–1067 in D. M. Pearsall, Ed., *Encyclopedia of Archaeology*. Amsterdam: Elsevier/Academic Press.

Bollen, Kenneth A. and David P. Phillips. 1981. Suicidal motor fatalities in Detroit: A replication. *American Journal of Sociology* **87**:404–412.

Bollen, Kenneth A. and David P. Phillips. 1982. Imitative suicides: A national study of the effects of television news stories. *American Sociological Review* **47**:802–809.

Bolnick, Daniel I. and Benjamin M. Fitzpatrick. 2007. Sympatric speciation: Models and empirical evidence. *Annual Review of Ecology, Evolution, and Systematics*. **38**:459–487.

Bonner, John Tyler. 1980. *The Evolution of Culture in Animals*. Princeton, N. J.: Princeton University Press.

Bonner, John Tyler. 2006. *Why Size Matters*. Princeton: Princeton University Press.

Bowles, Samuel. 2008. Conflict: Altruism's midwife. *Nature* **456**:326–327.

Boyd, Robert and Peter J. Richerson. 1985. *Culture and the Evolutionary Process*. Chicago: University of Chicago Press.

Boyd, Robert and Peter J. Richerson. 1992. How microevolutionary processes give rise to history. Pp. 179–210 in Matthew H. and Doris V. Niteki, Eds., *History and Evolution*. State University of New York Press.

Boyd, Robert, Peter J. Richerson, Monique Borgerhoff-Mulder and William H. Durham. 1997. Are cultural phylogenies possible? Pp. 355–386 in P. Weingart, P. J. Richerson, S. D. Mitchell and S. Maasen, Eds., *Human by Nature: Between Biology and the Social Sciences*. Mahwah, N. J.: Lawrence Erlbaum Associates.

Brantingham, P. Jeffrey. 2007. A unified evolutionary model of archaeological style and function based on the price equation. *American Antiquity* **72**:395–416.

Brickman, Phillip, Dan Coates and Ronnie Janoff-Bulman. 1978. Lottery winners and accident victims: Is happiness relative? *Journal of Personality and Social Psychology* **36**:917–927.

Brockmann, Jane H. and Richard Dawkins. 1979. Joint nesting in a Digger Wasp as an evolutionarily stable preadaptation to social life. *Behaviour* **71**:203–245.

Brockmann, Jane H., Alan Grafen and Richard Dawkins. 1979. Evolutionarily stable nesting strategy in a Digger Wasp. *Journal of Theoretical Biology* **77**:473–496.

Brodie, Richard. 1996. *Virus of the Mind: The New Science of the Meme*. Seattle: Integral Press.

Brown, Jerram L. 1983. Cooperation – a biologist's dilemma. *Advances in the Study of Behavior* **13**:1–37.

Brown, P., T. Sutika, M. J. Morwood, R. P. Soejono, Jatmiko, E. Wayhu Saptomo and Rokus Awe Due. 2004. A new small-bodied hominin from the Late Pleistocene of Flores, Indonesia. *Nature* **431**:1055–1061.

Brumfiel, Geoff. 2008. Older scientists publish more papers. *Nature* **455**: 1161.

Buckley, William. 1967. *Sociology and Modern Systems Theory*. Englewood Cliffs, N. J.: Prentice-Hall Inc.

Burger, J., M. Kirchner, B. Bramanti, W. Haak and M. G. Thomas. 2007. Absence of the lactase-persistence-associated allele in early Neolithic Europeans. *Proceedings of the National Academy of Sciences, USA* **104**:3736–3741.

Burling, Robbins. 2005. *The Talking Ape: How Language Evolved*. New York: Oxford University Press.

Burnet, Macfarlane. 1968. *Changing Patterns*. Melbourne: William Heinemann Ltd.

Burt, Austin and Robert Trivers. 2006. *Genes in Conflict: The Biology of Selfish Genetic Elements*. Cambridge: The Belknap Press of Harvard University Press.

Buss, Leo W. 1987. *The Evolution of Individuality*. Princeton: Princeton University Press.

Caldwell, Roy and David R. Lindberg. 2009. *Understanding Evolution*. University of California Museum of Paleontology. http://evolution.berkeley.edu/evolibrary/article/O_O_O/evo_03

Callebaut, Werner and Diego Rassskin-Gutman, Eds., 2005. *Modularity: Understanding the Development and Evolution of Natural Complex Systems*. Cambridge: The MIT Press.

Campbell, Donald T. 1960. Blind variation and selective retention in creative thought as in other knowledge processes. *Psychological Review* **67**:380–400.

Campbell, Donald T. 1965. Variation and selective retention in socio-cultural evolution. Pp. 19–49 in H.R. Barringer, G.I. Blanksten and R.W. Mack, Eds., *Social Change in Developing Areas*. Cambridge: Schenkman Press.

Campbell, Donald T. 1974. Evolutionary epistemology. Pp. 349–374 in Paul Arthur Schilpp, Ed., *The Philosophy of Karl R. Popper*. La Salle, Ill.: Open Court.

Carroll, Sean B. 2005. *Endless Forms Most Beautiful: The New Science of Evo Devo*. New York: W.W. Norton & Company, Inc.

Carroll, Sean B. 2006. *The Making of the Fittest: DNA and the Ultimate Forensic Record of Evolution*. New York: W.W. Norton & Company, Inc.

Carroll, Sean B., Jen Grenier and Scott Wetherbee. 2000. *From DNA to Diversity*. Oxford: Blackwell Science.

Casti, John L. and Anders Karlqvist, Eds. 1995. *Cooperation and Conflict in General Evolutionary Processes*. New York: Wiley.

Catchpole, Clive K. and Peter J.B. Slater. 2008. *Bird Song: Biological Themes and Variations*. Cambridge: Cambridge University Press.

Cavalli-Sforza, Luigi Luca. 1997. Genes, people and languages. *Proceedings of the National Academy of Sciences USA* **94**:7719–7724.

Cavalli-Sforza, Luigi Luca. 2000. *Genes, People and Languages*. Berkeley: University of California Press.

Cavalli-Sforza, Luigi L. and Marcus W. Feldman. 1981. *Cultural Transmission and Evolution: A Quantitative Approach*. Princeton: Princeton University Press.

Chapais, Bernard. 2008. *Primeval Kinship: How Pair-Bonding Gave Birth to Human Society*. Cambridge: Harvard University Press.

Charnov, Eric L. 1982. *The Theory of Sex Allocation*. Princeton: Princeton University Press.

Charnov, Eric L. 1993. *Life History Invariants: Some Explanations of Symmetry in Evolutionary Ecology*. Oxford: Oxford University Press.

Check, Erika. 2007. More biologists but tenure stays static. *Nature* **448**:848–849.

Choi, Jung-Kyoo and Samuel Bowles. 2007. The coevolution of parochial altruism and war. *Science* **318**:636–640.

Chomsky, Noam. 1957. *Syntactic Structures*. The Hague: Mouton.

Chomsky, Noam. 1959. Review of Verbal Behavior by B.F. Skinner. *Language* **35**:26–57.

Christakis, Nicholas A. and James H. Fowler. 2007. The spread of obesity in a large social network over 32 years. *The New England Journal of Medicine* **357**:370–379.

Christakis, Nicholas A. and James H. Fowler. 2008. The collective dynamics of smoking in a large social network. *The New England Journal of Medicine* **358**:2249–2258.

Christiansen, Morten H. and Simon Kirby, Eds., 2003. *Language Evolution*. Oxford University Press.

Clark, Samuel. 2005. Human intentionality in the functionalist theory of social change: The role of French provincial intendants in state–society differentiation. *European Journal of Sociology* **46**:213–262.

Clark, Terry, Ed. 1969. *On Communication and Social Influence*. Chicago: University of Chicago Press.

Clutton-Brock, Tim. 2007. Sexual selection in males and females. *Science* **318**:1882–1885.

Clutton-Brock, Tim. 2009. Sexual selection in females. *Animal Behavior* **77**:3–11.

Clutton-Brock, Tim H. and Geoffrey A. Parker. 1992. Potential reproductive rates and the operation of sexual selection. *The Quarterly Review of Biology* **67**:437–456.

Coase, Ronald K. 1937. The nature of the firm. *Economica* **4**:386–405.

Cochran, Gregory and Henry Harpending. 2009. *The 10,000 year Explosion: How Civilization Accelerated Human Evolution*. New York: Basic Books.

Cohen, Lawrence E., and Richard Machalek. 1988. A general theory of expropriative crime: An evolutionary ecological perspective. *American Journal of Sociology* **94**:465–501.

Cole, Stephen. 1992. *Making Science: Between Nature and Society*. Cambridge: Harvard University Press.

Coleman, James S. 1990. *Foundations of Social Theory*. Cambridge: The Belknap Press of Harvard University Press.

Collins, Harry M. 1985. *Changing Order: Replication and Induction in Scientific Practice*. London: Sage Publications.

Collins, Harry M. 2009. Essay: We cannot live by scepticism alone. *Nature* **485**:30–31.

Comte, Auguste. 1855, 1974. *The Positive Philosophy*. New York: AMS Press.

Conway-Morris, Simon. 2003. *Life's Solution: Inevitable Humans in a Lonely Universe*. Cambridge: Cambridge University Press.

Coolen, Isabelle, Yfke van Bergen, Rachel L. Day, and Kevin N. Laland. 2003. Species differences in adaptive use of public information in sticklebacks. *Proceeedings of the Royal Society of London B* **270**:2413–2419.

Corning, Peter A. 1982. Durkheim and Spencer. *British Journal of Sociology* **33**:359–82.

Corning, Peter A. 2005. *Holistic Darwinism: Synergy, Cybernetics and the Bioeconomics of Evolution*. Chicago: University of Chicago Press.

Covington, M.A., H. Congzhou, C. Brown *et al.* 2005. Schizophrenia and the structure of language: The linguists's view. *Schizophrenia Research* **77**:85–98.

Crichton, Michael. 2002. *Prey*. HarperCollins.

Crick, Francis H. C. 1968. The origin of the genetic code. *Journal of Molecular Biology* **38**:367–379.

Croft, William. 2000. *Explaining Language Change: An Evolutionary Approach*. Pearson Education Limited.

Croft, William. 2008. Evolutionary linguistics. *Annual Review of Anthropology* **37**:219–234.

Cronin, Helena. 1991. *The Ant and the Peacock – Altruism and Sexual Selection from Darwin to Today*. New York: Cambridge University Press.

Cullen, Ben Sanford. 2000. *Contagious Ideas: On Evolution, Culture, Archaeology, and Cultural Virus Theory*. London: Oxbow Books.

Cziko, Gary. 1995. *Without Miracles: Universal Selection Theory and the Second Darwinian Revolution*. Cambridge: MIT Press.

Danchin, Étienne, Luc-Alain Giraldeau, Thomas J. Valone and Richard H. Wagner. 2004. Public information: From nosy neighbors to cultural evolution. *Science* **305**:487–491.

Danchin, Étienne, Luc-Alain Giraldeau and Frank Cézilly. 2008. *Behavioural Ecology*. Oxford: Oxford University Press.

Davidson, Eric H. 2006. *The Regulatory Genome*. Burlington, MA: Academic Press.

Dawkins, Richard. 1976. *The Selfish Gene*. Oxford: Oxford University Press.

Dawkins, Richard. 1982. *The Extended Phenotype: The Gene as the Unit of Selection*. New York: W. H. Freeman and Company Limited.

Dawkins, Richard. 1995. *River Out of Eden: A Darwinian View of Life*. London: Weidenfeld & Nicolson.

Dawkins, Richard. 1996. *Climbing Mount Improbable*. New York: W. W. Norton & Company, Inc.

Dawkins, Richard. 2003. *A Devil's Chaplain: Reflections on Hope, Lies, Science and Love*. Boston: Houghton Mifflin Co.

Deacon, Terrace. 1997. *The Symbolic Species: The Co-evolution of Language and the Brain*. New York: W. W. Norton.

De Block, Andreas and Bart Du Laing. 2007. Paving the way for an evolutionary social constuctivism. *Biological Theory* **2**:337–348.

Dennett, Daniel C. 1995. *Darwin's Dangerous Idea: Evolution and the Meanings of Life*. New York: Simon & Schuster.

de Saussure, Ferinand. 1916, 1966. *Course in General Linguistics*. Translated by Wade Baskin. New York: McGraw-Hill.

de Waal, Frans. 1998. *Chimpanzee Politics: Power & Sex Among the Apes*. Baltimore: John Hopkins University Press.

de Waal, Frans. 2001. *The Ape and the Sushi Master: Cultural Reflections of a Primatologist*. New York: Basic Books.

Distin, Kate. 2005. *The Selfish Meme*. Cambridge: Cambridge University Press.

Dobzhansky, Theodosius. 1973. *The American Biology Teacher* **35**:125–129.

Doing, Park. 2008. Give me a laboratory and I will raise a discipline: The past, present and future politics of laboratory studies in STS. Pp. 279–295 in Edward J. Hackett, Olga Amsterdamska, Michael Lynch and Judy Wajcman, Eds., *The Handbook of Science and Technology Studies*. Cambridge: The MIT Press.

Domes, Katja, Roy A. Norton, Mark Maraun and Stefan Scheu. 2007. Reevolution of sexuality breaks Dollo's law. *Proceedings of the National Academy of Sciences USA* **104**:7139–7144.

Donahoe, John H., and Rocío Vegas. 2004. Pavlovian conditioning: The cs-ur relation. *Journal of Experimental Psychology: Animal Behavior Processes* **30**:17–33.

Driesch, Hans. 1914. *The History and Theory of Vitalism*. London: Macmillan and Company Limited.

Drori, Gili S. 2003. *Science in the Modern World Polity: Institutionalization and Globalization*. Stanford: Stanford University Press.

Dugatkin, Lee. 1997. *Cooperation Among Animals: An Evolutionary Perspective*. New York: Oxford University Press.

Dugatkin, Lee 1999. *Cheating Monkeys and Citizen Bees: The Nature of Cooperation in Animals and Humans*. New York: The Free Press.

Dugatkin, Lee Alan. 2000. *The Imitation Factor: Evolution Beyond the Gene*. New York: The Free Press.

Dunn, Jennifer. 2006. When ontologies collide it's a slippery slope: Differing representations of the social construction of child sexual abuse. *Contemporary Sociology* **35**:355–358.

Dunnell, Robert C. 1971. *Systematics in Prehistory*. New York: The Free Press.

Durand, Rodolphe. 2006. *Organizational Evolution and Strategic Management*. London: Sage Publications Ltd.

Durham, William H. 1991. *Coevolution: Genes, Culture and Human Diversity*. Stanford: Stanford University Press.

Durkheim, Emile. 1893, 1964. *The Division of Labor in Society*. Glencoe, Ill.: The Free Press.

Durkheim, Emile. 1895, 1982. *The Rules of Sociological Method*. New York: The Free Press.

Durkheim, Emile. 1897, 2006. *On Suicide*. London: Penguin.

Durkheim, Emile. 1912, 1965. *The Elementary Forms of the Religious Life*. Glencoe, Ill.: The Free Press.

Duthie, A. Bradley. 2004. The fork and the paperclip: A memetic perspective *Journal of Memetics* **8**:3Pp.

Ebach, Malte C., Juan J. Morrone and David M. Williams. 2008. A new cladistics of cladists. *Biology and Philosophy* **23**:153–156.

Eberhard, William G. 1996. *Female Control: Sexual Selection by Cryptic Female Choice*. Princeton, N. J.: Princeton University Press.

Edelman, Gerald. 1987. *Neural Darwinism*. New York: Basic Books.

Ehrlich, Paul R. 2000. *Human Natures: Genes, Cultures, and the Human Prospect*. Washington: Island Press/Shearwater Books.

Elinson, Richard P. 1987. Change in developmental patterns: Embryos of amphibians with large eggs. Pp. 1–21 in Rudolf A. Raff and Elizabeth C. Raff, Eds., *Development as an Evolutionary Process*. New York: Alan R. Liss, Inc.

Emlen, Stephen T. and Lewis W. Oring. 1977. Ecology, sexual selection, and the evolution of mating systems. *Science* **197**:215–223.

Engel, Jonathan. 2006. *The Epidemic: A Global History of Aids*. New York: Harpercollins Publishers.

EOL. 2008-9. *Encyclopedia of Life*. Open source. http://eol.org

Essinger, James. 2004. *Jacquard's Web: How a Hand Loom Led to the Birth of the Information Age*. Oxford: Oxford University Press.

Evans, Nicholas and Stephen Levinson. 2009. The myth of language universals: Language diversity and its importance for cognitive science. Forthcoming, *Behavioral and Brain Sciences*.

Everett, Daniel L. 2005. Cultural constraints on grammar and cognition in Pirahã. *Current Anthropology* **46**:621–646.

Fairbairn, Daphne J., Wolf U. Blanckenhorn and Tamás Székely, Eds., 2007. *Sex, Size, and Gender Roles: Evolutionary Studies of Sexual Size Dimorphism*. Oxford: Oxford University Press.

Falk, Dean, Charles Hildebolt, Kirk Smith *et al.* 2005. The brain of LB1, *Homo floresiensis. Science* **308**:242–245.

Fehr, Ernst and Herbert Gintis. 2007. Human motivation and social cooperation: Experimental and analytical foundations. *Annual Review of Sociology* **33**:43–74.

Feldman, Marcus W. and Kevin N. Laland. 1996. Gene-culture coevolutionary theory. *Trends in Ecology and Evolution* **11**:453–457.

Felsenstein, Joseph. 2004. *Inferring Phylogenies*. Sinauer Associates.

Felsenstein, Joseph. 2009. *Phylogeny Programs*. National Science Foundation. http://evolution.genetics.washington.edu/phylip/software.html.

Feyerabend, Karl. 1975. *Against Method: Outline of a Theory of Knowledge*. London: NLB.

Feyerabend, Karl. 1978. *Science in a Free Society*. London: NLB.

Field, Alexander J. 2001. *Altruistically Inclined? The Behavioral Sciences, Evolutionary Theory, and the Origins of Reciprocity*. Ann Arbor: The University of Michigan Press.

Fitze, Patrick S. and Jean-François Le Galliard. 2008. Operational sex ratio, sexual conflict and the intensity of sexual selection. *Ecology Letters* **11**:432–439.

Fitzpatrick, Benjamin M., J. Fordyce and S. Gavrilets. 2008. What, if anything, is sympatric speciation? *Journal of Evolutionary Biology* **21**:1452–1459.

Fog, Agner. 1997. Cultural r/K selection. *Journal of Memetics* **1**. http://jom-emit. cfpm.org/

Fog, Agner. 1999. *Cultural Selection*. Dordrecht: Kluwer Academic Publishers.

Foley, Jonathan A., Ruth DeFries, Gregory P. Asner *et al.* 2005. Global consequences of land use. *Science* **309**:570–574.

Foley, Jonathan A., Chad Monfreda, Navin Ramankutty and David Zaks. 2007. Our share of the planetary pie. *Proceedings of the National Academy of Sciences USA* **104**:12585–12586.

Folger, Tim. 2008. A universe built for us. *Discover* **29**(12):52–58.

Fortunato, Laura. 2008. "A phylogenetic approach to the history of cultural practices." Pp. 189–199 in Nicholas J. Allen, Hilary Callan, Robin Dunbar and Wendy James, Eds. *Early Human Kinship: From Sex to Social Reproduction*. Oxford: Blackwell Publishing Ltd.

Foster, David J. and Matthew A. Wilson. 2006. Reverse replay of behavioural sequences in hippocampal place cells during the awake state. *Nature* **440**:680–683.

Fowler, James H. and Nicholas A. Christakis. 2008. Dynamic spread of happiness in a large social network: Longitudinal analysis over 20 years in the Framingham Heart Study. *British Medical Journal* **337**:a2338 doi:10.1136/bmj.a2338.

Fracchia, Joseph and Richard C. Lewontin. 1999. Does culture evolve? *History and Theory* **38**:52–78.

Fracchia, Joseph and Richard C. Lewontin. 2005. The price of metaphor. *History and Theory* **44**:14–29.

Frank, Steven A. 1998. *Foundations of Social Evolution*. Princeton, N. J.: Princeton University Press.

Frankino, W. Anthony, Bas J. Zwaan, David L. Stern and Paul M. Brakefield. 2007. Internal and external constraints in the evolution of morphological allometries in a butterfly. *Evolution* **61**:2958–2970.

Fudenberg, Drew and David K. Levine. 1998. *The Theory of Learning in Games*. Cambridge: MIT Press.

Fuller, Steve. 2006. *The Philosophy of Science and Technology Studies*. New York: Routledge.

Fuller, Steve. 2008. Science studies goes public: a report on ongoing performance. *Spontaneous Generations: A Journal for the History and Philosophy of Science* **1** (2):11–21.

Futuyma, Douglas J. 1998. *Evolutionary Biology*, 3rd edn. Sinauer Associates, Inc.

Gadagkar, Raghavendra. 1997. *Survival Strategies: Cooperation and Conflict in Animal Societies*. Cambridge: Havard University Press.

Galbraith, John Kenneth. 1958. *The Affluent Society*. Boston: Houghton Mifflin.

Galbraith, John Kenneth. 1967. *The New Industrial State*. Boston: Houghton Mifflin.

Galef, Bennett G. 1992. The question of animal culture. *Human Nature* **3**:157–178.

Gallese, Vittorio, Luciano Fadiga, Leonardo Fogassi and Giacomo Rizzolatti. 1996. Action recognition in the premotor cortex. *Brain* **119**:593–609.

Garfinkel, Harold. 1967. *Studies in Ethnomethodology*. Englewood Cliffs, N. J.: Prentice-Hall.

Gerard, R. W., Clyde Kluckhohn and Anatol Rapoport. 1956. Biological and cultural evolution: some analogies and explorations. *Behavioral Science* **1**:6–34.

Ghiselin, Michael. 1974a. Comment on D. Freeman's The evolutionary theories of Charles Darwin and Herbert Spencer. *Current Anthropology* **15**:224.

Ghiselin, Michael. 1974b. A radical solution to the species problem. *Systematic Zoology* **23**:536–544.

Giddens, Anthony. 1976. *New Rules of Sociological Method: A Positive Critique of Interpretive Sociologies*. London: Hutchinson.

Giddens, Anthony. 1979. *Central Problems in Social Theory: Action, Structure and Contradiction in Social Analysis*. Berkeley: University of California Press.

Giddens, Anthony. 1984. *The Constitution of Society: Outline of the Theory of Structuration*. Polity Press.

Gilbert, R. M. 1970. Psychology and biology. *The Canadian Psychologist.* **12**:221–238.

Gilbert, Richard M. 1972. Variation and selection of behavior. Pp. 264–276 in Richard M. Gilbert and John R. Millenson, Eds., *Reinforcement: Behavioral Analyses*. New York: Academic Press.

Gintis, Herbert. 2007. A framework for the unification of the behavioral sciences. *Behavioral and Brain Sciences* **30**:1–61.

Gintis, Herbert, Samuel Bowles, Robert T. Boyd and Ernst Fehr, Eds., 2005. *Moral Sentiments and Material Interests: The Foundations of Cooperation in Economic Life*. Cambridge: MIT Press.

Glenn, Sigrid S. 1991. Contingencies and metacontingencies: Relations among behavioral, cultural, and biological evolution. Pp. 39–73 in Peter A. Lamal, Ed., *Behavioral Analysis of Societies and Cultural Practices*. Hemisphere Press.

Glenn, Sigrid S., Janet Ellis and Joel Greenspoon. 1992. On the revolutionary nature of the operant as a unit of behavioral selection. *American Psychologist* **47**:1329–1336.

Glenn, Sigrid S. and D. P. Field. 1994. Functions of the environment in behavioral evolution. *The Behavior Analyst* **17**:241–259.

Glenn, Sigrid S. and G. J. Madden. 1995. Units of interaction, evolution and replication: Organic and behavioral parallels. *The Behavior Analyst* **18**:237–251.

Godfrey-Smith, Peter. 2007. Information in biology. Pp. 103–119 in David Hull and Michael Ruse, Eds., *The Cambridge Companion to the Philosophy of Biology*. Cambridge: Cambridge University Press.

Goldstein, E. Bruce. 2007. *Sensation and Perception*. Australia: Thomas /Wadsworth.

Goldstein, E. Bruce. 2008. *Cognitive Psychology: Connectng Mind, Research, and Everyday Experience*, 2nd edn. Thomson Wadsworth.

Gould, Stephen Jay. 1980. A biological homage to Mickey Mouse. Pp. 95–107 in Stephen Jay Gould, *The Panda's Thumb: More Reflections on Natural History*. New York: W. W. Norton & Company.

Gould, Stephen J. 1981, 1996. *The Mismeasure of Man*. New York: W. W. Norton & Co., Inc.

Gould, Stephen J. 1989. *Wonderful Life: The Burgess Shale and the Nature of History*. New York: W. W. Norton & Company, Inc.

Gould, Stephen J. 2002. *The Structure of Evolutionary Theory*. Cambridge, MA.: The Belknap Press of Harvard University Press.

Grafen, Alan. 1991. Modellling in behavioural ecology. Pp. 5–31 in John R. Krebs and Nicholas B. Davies, Eds., *Behavioural Ecology: An Evolutionary Approach*. Third edition. Oxford: Blackwell Scientific Publications.

Grant, Peter R. and B. Rosemary Grant. 2007. *How and Why Species Multiply: The Radiation of Darwin's Finches*. Princeton: Princeton University Press.

Gray, D. Russell and Quentin D. Atkinson. 2003. Language-tree divergence times support the Anatonian theory of Indo-European Origin. *Nature* **426**:435–439. (See also commentary by David B. Searls, pp. 391–392.)

Gray, D. Russell and Fiona M. Jordan. 2000. Language trees support the express-train sequence of Austronesian expansion. *Nature* **405**:1052–1055. (See also commentary by Rebecca L. Cann, pp. 1008–1009.)

Gray, D. Russell, Simon J. Greenhill, and Robert M. Ross. 2007. The pleasures and perils of darwinizing culture (with phylogenies). *Biological Theory* **2**:360–375.

Greenberg, Daniel S. 2007. *Science for Sale: The Perils, Rewards and Delusions of Campus Capitalism*. Chicago: The University of Chicago Press.

Greene, Mott. 2007. The demise of the lone author. *Nature* **450**:1165.

Greenpeace International. Aug. 17, 2005. Recycling of electronic wastes in India and China: Workplace & environmental contamination. http://www.green.peace.org/raw/content/international/press/reports/recycling-of-electronic-waste.pdf

Griesemer J. R. 2000. Reproduction and the reduction of genetics. Pp. 240–285 in P. J. Beurton, R. Falk and H. Rheinberger, Eds., *The Concept of the Gene in Development and Evolution: Historical and Epistemological Perspectives*. Cambridge: Cambridge University Press.

Griffiths, Paul E. 2002. Lost: one gene concept. Reward to finder. *Biology and Philosophy* **17**:271–283.

Güldemann, Tom and Mark Stoneking. 2008. A historical appraisal of clicks: a linguistic and genetic population perspective. *Annual Review of Anthropology* **37**:93–109.

Hacking, Ian. 1999. *The Social Construction of What?* Cambridge: Harvard University Press.

Hadany, Lilach and Josep M. Comeron. 2008. Why are sex and recombination so common? *Annals of the New York Academy of Sciences* **1133**:26–43.

Haggbloom, Steven J., Renee Warnick, James E. Warnick *et al*. 2002. The 100 most eminent psychologists of the 20th century. *Review of General Psychology* **6**:139–152.

Hall, Brian K. 1997. Phylotypic stage or phantom: Is there a highly conserved embryonic stage in vertebrates. *Trends in Ecology and Evolution* **12**:461–463.

Hall, Brian S. 2004. *Phylogenetic Trees Made Easy: A How-to Manual*. 2nd edn. Sinauer Associates.

Hall, John A. and Joseph M. Bryant. 2005. *Historical Methods in the Social Sciences*. Four Volumes. London: Sage Publications.

Hamilton, William D. 1964a. The genetical evolution of social behaviour, I. *Journal of Theoretical Biology* **7**:1–16.

Hamilton, William D. 1964b.The genetical evolution of social behaviour, II. *Journal of Theoretical Biology* **7**:17–52.

Hamilton, William D. 2001. *Narrow Roads of Gene Land, Volume II Evolution of Sex*. Oxford: Oxford University Press.

Hammerstein, Peter. 2003. *Genetic and Cultural Evolution of Cooperation*. Cambridge: The MIT Press.

Hansell, Michael H. 1984. *Animal Architecture and Building Behavior*. London: Longmans.

Hansell, Michael H. 2000. *Birds Nests and Construction Behaviour*. Cambridge, UK: Cambridge University Press.

Harris, Marvin. 1977. *Cannibals and Kings: The Origins of Cultures*. New York: Random House.

Hauser, Marc D., Noam Chomsky and W. Tecumseh Fitch. 2002. The faculty of language: what is it, who has it, and how did it evolve. *Science* **298**:1569–1579.

Hawken, Paul, Amory Lovins and L. Hunter Lovins. 1999. *Natural Capitalism: Creating the Next Industrial Revolution*. New York: Little, Brown and Company.

Hawks, John, Eric T. Wang, Gregory M. Cochran, Henry C. Harpending and Robert K. Moyzis 2007. Recent acceleration of human adaptive evolution. *Proceedings of the National Academy of Sciences USA* **104**:20753–20758.

Heckathorn, Douglas D. 2001. Sociological rational choice. Pp. 273–284 in George Ritzer and Barry Smart, Eds., *Handbook of Social Theory*. London: Sage Publications.

Hennig, Willi. 1966, *Phylogenetic Systematics*. Urbana: University of Illinois Press.

Herrnstein, Richard J. 1970. On the law of effect. *Journal of the Experimental Analysis of Behavior* **13**:243–266.

Herrnstein, Richard J. 1990. Rational choice theory: Necessary but not sufficient. *American Psychologist* **45**:356–367.

Herrnstein, Richard J. 1997. *The Matching Law: Papers in Psychology and Economics.* Cambridge: Harvard University Press.

Hewitt, John P. 2006. *Self and Society: A Symbolic Interactionist Social Psychology,* 10th edn. Boston: Allyn and Bacon.

Hickey, Donald A. and Michael R. Rose. 1988. The role of gene transfer in the evolution of eukaryotic sex. Pp. 161–175 in Richard E. Michod and Bruce R. Levin, Eds., *The Evolution of Sex: An Examination of Current Ideas.* Sunderland, Mass.: Sinauer Associates Inc.

Hill, William G., Michael E. Goddard and Peter M. Visscher. 2008. Data and theory point to mainly additive genetic variance for complex traits. *PloS Genetics* **4**:e1000008.

Hinde, Robert A., and Les A. Barden. 1985. The evolution of the teddy bear. *Animal Behaviour* **33**:1371–1373.

Hodgson, Geoffrey M. 1999. *Evolution and Institutions: On Evolutionary Economics and the Evolution of Economics.* Cheltenham, UK: Edward Elgar Publishing.

Hodgson, Geoffrey M. 2001. *How Economics Forgot History: The Problem of Historical Specificity in Social Science.* London: Routledge.

Hodgson, Geoffrey M. 2004. *The Evolution of Institutional Economics: Agency, Structure and Darwinism in American Institutionalism,* London and New York: Routledge.

Hodgson, Geoffrey M. and Thørbjorn Knudsen. 2006. Dismantling Lamarckism: Why descriptions of socio-economic evolution as Lamarckian are misleading. *Journal of Evolutionary Economics* **16**:343–366.

Holden, Clare J., and Ruth Mace. 2003. Spread of cattle led to the loss of matrilineal descent in Africa: A coevolutionary hypothesis. *Proceedings of the Royal Society of London* B **270**:2425–2433.

Holden, Clare J., and Ruth Mace. 2005. The cow is the enemy of matriliny: Using phylogenetic methods to investigate cultural evolution in Africa. Pp. 217–234 in Ruth Mace, Clare J. Holden and Stephen Shennan, Eds., *The Evolution of Cultural Diversity: A Phylogenetic Approach.* London: UCL Press.

Holmes, Frederic L. 2000. Seymour Benzer and the definition of the gene. Pp. 115–155 in Peter J. Beurton, Ralph F. Falk and Hans-Jörg Rheinberger, Eds., *The Concept of the Gene in Development and Evolution: Historical and Epistemological Perspectives.* Cambridge: Cambridge University Press.

Homans, George. 1961, 1974. *Social Behavior: Its Elementary Forms.* New York: Harcourt Brace and World.

Homans, George. 1987. Review of James S. Coleman's *Individual Interests and Collective Action: Selected Essays. Contemporary Sociology* **16**:769–770.

Hoppitt, Will and Kevin N. Laland. 2008. Social processes influencing learning in animals: A review of the evidence. *Advances in the Study of Behavior* **38**:105–165.

Hoppitt, Will, G. R. Brown, R. Vendall, A. Thornton, M. J. Webster and K. N. Laland. 2008. Lessons from animal teaching. *Trends in Ecology and Evolution* **23**:486–493.

Howard, Maureen L. and K. Geoffrey White. 2003. Social influence in pigeons (Columbia Livia): The role of differential reinforcement. *Journal of the Experimental Analysis of Behavior* **79**:175–191.

Huff, Toby E. 1993. *The Rise of Early Modern Science.* Cambridge: Cambridge University Press.

Hull, David L. 1978a. A matter of individuality. *Philosophy of Science* **45**:335–360.

Hull, David L. 1978b. Altruism in science: A sociobiological model of co-operative behavior among scientists. *Animal Behavior* **26**:685-697.

Hull, David L. 1988. *Science as a Process. An Evolutionary Account of the Social and Conceptual Development of Science.* Chicago: University of Chicago Press.

Hull, David L. 1994. Natural truths. *Nature* **368**:504-5.

Hull, David L. 2001a. *Science and Selection: Essays on Biological Evolution and the Philosophy of Science.* Cambridge: Cambridge University Press.

Hull, David L. 2001b. Review of John Ziman's *Real Science: What It Is, and What It Means. Nature* **411**:134-135.

Hull, David L., Rodney E. Langman, and Sigrid S. Glenn. 2001. A general account of selection: biology, immunology and behavior. *Behavioral and Brain Sciences* **24**:511-573.

Hunt, G. R. and R. D. Gray. 2003. Diversification and cumulative evolution in New Caledonian crow tool manufacture. *Proceedings of the Royal Society London B* **270**:867-874.

Hurley, Susan L. and Nick Chater. 2005. *Perspectives on Imitation: From Neuroscience to Social Science. VI Mechanisms of Imitation and Imitation in Animals, VII Imitation, Human Development, and Culture.* Cambridge: MIT Press.

Hurst, Laurence D. 2009. A positive becomes a negative. *Nature* **2009**:543-544.

Hurt, Teresa D., and Gordon F.M. Rakita. 2001. *Style and Function: Conceptual Issues in Evolutionary Archaeology.* Westport, Conn.: Bergin & Garvey.

Inglis, David and John Hughson. 2003. *Confronting Culture: Sociological Vistas.* Cambridge: Polity Press.

Iyer, Priya and Joan Roughgarden. 2008. Genetic conflict versus contact in the evolution of anisogamy. *Theoretical Population Biology* **73**:461-472.

Jablonka, Eva and Marion J. Lamb. 2006. *Evolution in Four Dimensions.* Cambridge: The MIT Press.

Jan, Steven. 2003. The evolution of a 'memeplex' in late Mozart: replicated structures in Pamina's 'Ach ich fühl's. *Journal of the Royal Musical Assocation* **128**:30-70.

Jan, Steven. 2007. *The Memetics of Music: A Neo-Darwinian View of Musical Structure and Culture.* Aldershot, England: Ashgate Publishing Limited.

Jenkins, Peter F. 1978. Cultural transmission of song patterns and dialect development in a free-living bird population. *Animal Behaviour* **27**:50-78.

Jerne, Niels Kai. 1967. Antibodies and learning: Selection versus instruction. Pp. 200-205 in Gardner C. Quarton, Theodore Melnechuk and Francis O. Schmitt, Eds., *The Neurosciences: A Study Programme.* Rochester: Rochester University Press.

Joly-Mascheroni, Ramiro M., Atsuchi Senju, and Alex J. Shepherd. 2008. Dogs catch human yawns. *Proceedings of the Royal Society B: Biology Letters* **4**:446-448.

Joy, Bill. 2000. Why the future doesn't need us. *Wired* Issue 8.04, April.

Judson, Olivia. 2002. *Dr. Tatiana's Sex Advice to all Creation: The Definitive Guide to the Evolutionary Biology of Sex.* New York: Metropoloitan Books of Henry Holt and Company.

Jungers, W.L., W.E.H. Harcourt-Smith, R.E. Wunderlich, M.W. Tocheri, S.G. Larson, T. Sutikna, Rhodus Awe Due and M.J. Morwood. 2009. The foot of Homo floresiensis. *Nature* **459**:81-84.

Kahneman, Daniel and Amos Tversky. 1979. Prospect theory: An analysis of decision under risk. *Econometrica* **47**:263-291.

Keller, Laurent, Ed. 1999. *Levels of Selection in Evolution.* Princeton, N. J.: Princeton University Press.

Kirschner, Marc W. and John C. Gerhart. 2005. *The Plausibility of Life: Resolving Darwin's Dilemma.* New Haven: Yale University Press.

Klüver, Jürgen. 2002. *An Essay Concerning Sociocultural Evolution: Theoretical Principles and Mathematical Models*. Dordrecht: Kluwer Academic Publishers.

Klüver, Jürgen. 2003. Historical evolution and mathematical models: A sociocultural algorithm. *Journal of Mathematical Sociology* **27**:53–83.

Knight, Frank H. 1921, 1971. *Risk, Uncertainty and Profit*. Chicago: University of Chicago Press.

Knight, Mairi E. and George F. Turner. 2004. Laboratory mating trials indicate incipient speciation by sexual selection among populations of the cichlid fish *Pseudotropheus zebra* from Lake Malawi. *Proceedings of the Royal Society of London B* **271**:675–680.

Knight, Rob. 2007. Reports of the death of the gene are greatly exaggerated. *Biology and Philosophy* **22**:293–306.

Knorr Cetina, Karin. 1981. *The Manufacture of Knowledge: An Essay on the Constructivist and Contextual Nature of Science*. Oxford: Pergamon Press.

Knowles, David. 1963. *The Monastic Order in England*. Cambridge: Cambridge University Press.

Kokko, Hanna, Michael D. Jennions and Robert Brooks. 2006. Unifying and testing models of sexual selection. *Annual Review of Ecology Evolution and Systematics* **37**:43–66.

Kroeber, Alfred L. and Talcott Parsons. 1958. The concepts of culture and of social system. *American Sociological Review* **23**:582–583.

Kronfeldner, Maria E. 2006. Is cultural evolution Lamarckian? *Biology and Philosophy* **22**:493–512.

Laland, Kevin N. 2008. Animal cultures. *Current Biology* **18**:R366–R370.

Latour, Bruno and Steve Woolgar. 1979. *Laboratory Life: The Social Construction of Scientific Facts*. London: Sage Publications.

Lecointre, Guillaume and Herve Le Guyader. 2006. *The Tree of Life: A Phylogenetic Classification*. Cambridge, M.A.: President and Fellows of Harvard College.

Lenski, Gerhard E. 1984. *Power and Privilege: A Theory of Social Stratification*. Chapel Hill: University of North Carolina Press.

Lenski, Gerhard. 2005. *Ecological-Evolutionary Theory: Principles and Applications*. Boulder: Paradigm Publishers.

Levins, Richard. 1968. *Evolution in Changing Environments*. Princeton: Princeton University Press.

Lévi-Strauss, Claude. 1969. *The Elementary Structures of Kinship*. Boston: Beacon Press.

Lewontin, Richard C., Steven Rose and Leon J. Kamin. 1984. *Not in Our Genes: Biology, Ideology and Human Nature*. New York: Pantheon.

Lieberson, Stanley. 2000. *A Matter of Taste: How Names, Fashion, and Culture Changes*. New Haven: Yale University Press.

Liesen, Laurette T. 2007. Women, behavior, and evolution. *Politics and the Life Sciences* **26**:51–70.

Lindenberg, Siegwart. 1990. Homo socio-economicus: The emergence of a general model of man in the social sciences. *Journal of Institutional and Theoretical Economics* **146**:727–748.

Lindenberg, Siegwart. 2001. Social rationality versus rational egoism. Pp. 635–668 in Jonathan H. Turner, Ed., *Handbook of Sociological Theory*. New York: Kluwer Academic/Plenum Publishers.

Lipo, Carl P., Michael J. O'Brien, Mark Collard and Stephen K. Shannon, Eds., 2006. *Mapping Our Ancestors: Phylogenetic Approaches in Anthropology and Prehistory*. New Brunswick, N. J.: Aldine Transaction.

Lodish, Harvey, Arnold Berk, S. Lawrence Zipursky, Paul Matsudaira, David Baltimore and James Darnell. 2000. *Molecular Cell Biology*. New York: W. H. Freeman and Company.

Loewenstein, Yonatan and H. Sebastian Seung. 2006. Operant matching is a genetic outcome of synaptic plasticity based on the covariance between reward and neural activity. *Proceedings of the National Academy of Sciences* **103**:15 224–15 229.

Lorenz, Konrad. 1967. *On Aggression*. New York: Bantam Books.

Lovejoy, Arthur O. 1936. *The Great Chain of Being: A Study of the History of an Idea*. Cambridge: Harvard University Press.

Luhmann, Niklas. 1995. *Social Systems*. Stanford: Stanford University Press.

Lumsden, Charles J. and Edward O. Wilson. 1981. *Genes, Mind and Culture: The Coevolutionary Process*. Cambridge: Harvard University Press.

Lynch, Aaron. 1996. *Thought Contagion: How Belief Spreads through Society*. New York: Basic Books.

Lynch, A., G. M. Plunkett, A. J. Baker, and P. F. Jenkins. 1989. A model of cultural evolution of chaffinch song derived with the meme concept. *The American Naturalist* **133**:634–653.

Lynch, Michael. 1985. *Art and Artifact in Laboratory Science: A Study of Shop Work and Shop Talk in a Research Laboratory*. London: Routledge & Kegan Paul.

MacArthur, Robert H. 1962. Some generalized theorems of natural selection. *Proceedings of the National Academy of Sciences U.S.A.* **48**:1893–1897.

MacArthur, Robert H. and Edward O. Wilson. 1967. *The Theory of Island Biogeography*. Princeton: Princeton University Press.

Mace, Ruth and Clare J. Holden. 2005. A phylogenetic approach to cultural evolution. *Trends in Ecology and Evolution* **20**:116–121.

Mace, Ruth, Clare J. Holden, and Stephen Shennan, Eds., 2005. *The Evolution of Cultural Diversity: A Phylogenetic Approach*. London: UCL Press.

Mace, Ruth and Fiona Jordan. 2005. The evolution of human sex ratio at birth: a bio-cultural analysis. Pp. 207–216 in Ruth Mace, Clare J. Holden and Stephen Shennan, Eds., *The Evolution of Cultural Diversity: A Phylogenetic Approach*. London: UCL Press.

Mace, Ruth and Mark Pagel. 1994. The comparative method in anthropology. *Current Anthropology* **35**:549–564.

Macy, Michael W. 2006. Rational choice. Pp. 70–87 in Peter J. Burke, Ed., *Contemporary Social Psychological Theories*. Stanford: Stanford University Press.

Mallon, Ron. 2009. Naturalistic approaches to social construction. *Stanford Encyclopedia of Philosophy*. http://plato.stanford.edu/entries/social-construction-naturalistic/

Marcot, Jonathan D. and Daniel W. McShea. 2007. Increasing hierarchical complexity throughout the history of life: phylogenetic tests of trend mechanisms. *Paleobiology* **32**:182–200.

Marcus, Gary. 2008. *Kluge: The Haphazard Construction of the Human Mind*. New York: Houghton Mifflin Company.

Marcus, Joyce. 2008. The archaeological evidence for social evolution. *Annual Review of Anthropology* **37**:251–266.

Marx, Karl. 1859, 1959. Excerpt from a conribution to the critique of political economy. Pp. 42–46 in Lewis S. Feuer, Ed., *Marx & Engels: Basic Writings on Politics and Philosophy*. New York: Doubleday & Company, Inc.

Math Bench. 2009. *Scaling*. University of Maryland. http://www.mathbench.umd.edu/mod207_scaling/page01.htm

Maynard Smith, John. 1978. *The Evolution of Sex*. Cambridge: Cambridge University Press.

Maynard Smith, John. 1982. *Evolution and the Theory of Games*. Cambridge: Cambridge University Press.

Maynard Smith, John. 1988. Mechanisms of advance. *Science* **242**:1182–1183.

Maynard Smith, John. 1998. *Shaping Life: Genes, Embryos and Evolution.* London: Weidenfeld & Nicolson, The Orion Publishing Group Ltd.

Maynard Smith, John and George R. Price. 1973. The logic of animal conflict. *Nature* **246**:15–18.

Maynard Smith, John and Eörs Szathmary. 1995. *The Major Transitions in Evolution.* Oxford: W. H. Freeman and Company Limited.

Maynard Smith, John and N. Warren. 1982. Models of cultural and genetic change. *Evolution* **36**:620–27.

Mayr, Ernst. 1942. *Systematics and the Origin of Species, From the Viewpoint of a Zoologist.* Cambridge: Harvard University Press.

McGhee, George. 2008. Convergent evolution: A periodic table of life? Pp. 17–31 in Simon Conway-Morris, Ed., *The Deep Structure of Biology: Is Convergence Sufficiently Ubiquitous to Give a Directional Signal?* West Conshohocken, Penn.: Templeton-Foundation Press.

McGrew, Wm. C. 1992. *Chimpanzee Material Culture: Implications for Human Evolution.* Cambridge: Cambridge University Press.

McGrew, Wm. C. 2003. Ten dispatches from the chimpanzee culture wars. Pp. 419–439 in Frans B. M. de Waal and Peter L. Tyack, Eds., *Animal Social Complexity: Intelligence, Culture, and Individualized Societies.* Cambridge: Harvard University Press.

McGrew, Wm. C. 2004. *The Cultured Chimpanzee: Reflections on Cultural Primatology.* Cambridge: Cambridge University Press.

McKelvey, Bill. 1982. *Organizational Systematics: Taxonomy, Evolution, Classification.* Berkeley: University of California Press.

McPherson, M., L. Smith-Lovin and J.M. Cook. 2001. Birds of a feather: Homophily in social networks. *Annual Review of Sociology* **27**:415–444.

Mead, George Herbert. In Charles W. Morris, Ed., 1934. *Mind, Self and Society.* Chicago: University of Chicago Press.

Merton, Robert K. 1942. The normative structure of science. In Norman W. Storer, Ed., Robert K. Merton. 1973. *The Sociology of Science: Theoretical and Empirical Investigations.* Chicago: University of Chicago Press.

Merton, Robert C. 1961. Singletons and multiples in scientific discovery. *Proceedings of the American Philosophical Society* **105**:470–486.

Merton, Robert K. 1968. Manifest and latent functions. In Robert K. Merton, Ed., *Social Theory and Social Structure.* New York: The Free Press.

Mesoudi, Alex, Andrew Whiten and Kevin N. Laland. 2004. Is human cultural evolution Darwinian? Evidence reviewed from the perspective of The Origin of Species. *Evolution* **58**:1–11.

Mesoudi, Alex, Andrew Whiten, and Kevin N. Laland. 2006. Towards a unified science of cultural evolution. *Behavioral and Brain Sciences* **29**:329–383.

Mikkola, Mari. 2008. Feminist perspectives on gender and sex. *Stanford Encyclopedia of Philosophy* http://plato.stanford.edu/entries/feminism_gender/

Miller, Neal E. and John Dollard. 1941, 1962. *Social Learning and Imitation.* New Haven: Yale University Press.

Mindell, David P. 2006. *The Evolving World: Evolution in Everyday Life.* Cambridge: Harvard University Press.

Molm, Linda D., Jessica L. Collett and David R. Schaefer. 2007. Building solidarity through generalized exchange: A theory of reciprocity. *American Journal of Sociology* **113**:205–242.

Molotch, Harvey. 2003. *Where Stuff Comes From.* New York and London: Routledge.

Monad, Jacques. 1971. *Chance and Necessity: An Essay on the Natural Philosophy of Modern Biology.* New York: Alfred A. Knopf, Inc.

Moran, Lawrence A. 2007. What is a gene? http://sandwalk.blogspot.com/2007/01/what-is-gene.html

Morris, Desmond. 1967. *The Naked Ape: A Zoologist's Study of the Human Animal*. New York: Crown.

Morris, Paul H., V. Reddy, and C. Bunting. 1995. The survival of the cutest: who's responsible for the evolution of the teddy bear. *Animal Behaviour* **50**:1697-1700.

Mufwene, Salikoko. 2001. *The Ecology of Language Evolution*. Cambridge: Cambridge University Press.

Mundinger, Paul C. 1980. Animal cultures and a general theory of cultural evolution. *Ethology and Sociobiology* **I**:183-223.

Murmann, Johann Peter *et al.* 2004. http://etss.net/index.php?/weblog/reference/debate_on_lamarckianism_in_social_evolution_june_2004/

Nagasato, Chikako. 2005. Behavior and function of paternally inherited centrioles in brown algal zygotes. *Journal of Plant Research* **118**:361-369.

Nettle, Daniel and Suzanne Romaine. 2000. *Vanishing Voices: The Extinction of the World's Languages*. New York: Oxford University Press.

Nishida, Toshida. 1987. Local traditions and cultural transmission. Pp. 462-474 in Barbara B. Smuts, D.L. Cheney, R.M. Seyfarth, R.W. Wrangham and T.T. Stuhsaker, Eds., *Primate Societies*. Chicago, Il.: University of Chicago Press.

Noë, Ronald and Peter Hammerstein. 1994. Biological markets: supply and demand determine the effect of partner choice in cooperation, mutualism and mating. *Behavioral Ecology and Sociobiology* **35**:1-11.

Noë, Ronald and Peter Hammerstein. 1995. Biological markets. *Trends in Ecology and Evolution* **10**:336-339.

Nowak, Martin A. 2006. Five rules for the evolution of cooperation. *Science* **314**:1560-1563.

O'Brien, Michael J. and R. Lee Lyman. 2000. *Applying Evolutionary Archaeology: A Systematic Approach*. New York: Kluwer Academic.

O'Brien, Michael J. and R. Lee Lyman. 2003. *Style, Function, Transmission: Evolutionary Archaeological Perspectives*. Salt Lake City: University of Utah Press.

O'Brien, Michael J., John Darwent, and R. Lee Lyman. 2001. Cladistics is useful for reconstructing archaeological phylogenies: Paleoindian points from the southeastern United States. *Journal of Archaeological Science* **28**:1115-1136.

O'Connell, Kevin F.O. 2000. The centrosome of the early *C. Elegans* embryo: Inheritance, assembly, replication, and developmental roles. Pp. 365-384 in Robert E. Palazzzo and Gerald P. Schatten, Eds., *The Centrosome in Cell Replication and Early Development*. New York: Academic Press.

Odling-Smee, F. John, Kevin N. Laland, and Marcus W. Feldman. 1996. Niche construction. *The American Naturalist* **146**:641-648.

Odling-Smee, F. John, Kevin N. Laland, and Marcus W. Feldman. 2003. *Niche Construction: The Neglected Process in Evolution*. Princeton: Princeton University Press.

O'Hara, Bob. 2008. A statistician wonders about the influence of additive variance. *Nature*: **452**:785.

O'Hara, Robert J. 1996. Trees of history in systematics and philology. *Memorie della Società Italiana di Scienze Naturali e del Museo Civico di Storia Naturale di Milano* **27**:81-88.

Ohno, Susumu. 1970. *Evolution by Gene Duplication*. Berlin: Springer-Verlag.

Ohtsuki, Hisashi, Yoh Iwasa and Martin A. Nowak. 2009. Indirect reciprocity provides only a narrow margin of efficiency for costly punishment. *Nature* **457**:79-81.

Okasha, Samir. 2006. *Evolution and the Levels of Selection*. Oxford: Oxford University Press.

Okasha, Samir. 2008. The units and levels of selection. Pp. 138–156 in Sahotra Sarkar and Anya Plutynski, Eds., *A Companion to the Philosophy of Biology*. Malden, MA: Blackwell Publishing Ltd.

Olson, Mancur. 1965, 1971. *The Logic of Collective Action: Public Goods and the Theory of Groups*. Cambridge: Harvard University Press.

Otto, S. P. and T. Lenormand. 2002. Resolving the paradox of sex and recombination. *Nature Reviews Genetics* **3**:252–261.

Otto, S. P. and A. C. Gerstein. 2006. Meiosis and the causes and consequences of recombination. *Biochemical Society Transactions* **34**:519–522.

Owings, Donald H. and Eugene S. Morton. 1998. *Animal Vocal Communication: A New Approach*. Cambridge: Cambridge University Press.

Oyama, Susan. 2000. *The Ontogeny of Information: Developmental Systems and Evolution*, 2nd edn. Durham: Duke University Press.

Pagel, Mark. 1994. Detecting correlated evolution on phylogenies: A general method for the comparative analysis of discrete characters. *Proceedings of the Royal Society of London B* **255**:37–45.

Parker, Geoffrey A. 1970. Sperm competition and its evolutionary consequences in the insects. *Biological Reviews* **45**:525–567.

Parker, Geoffrey. 1984. The producer/scrounger model and its relevance to sexuality. Pp. 127–152 in Christopher J. Barnard, Ed., *Producers and Scroungers: Strategies of Exploitation and Parasitism*. London: Croom Helm Ltd.

Parker, Geoffrey, R. Baker and V. Smith. 1972. The origin and evolution of gamete dimorphism and the male-female phenomenon. *Journal of Theoretical Biology* **36**:529–553.

Parsons, Talcott. 1966. *Societies: Evolutionary and Comparative Perspectives*. Englewood Cliffs, N. J.: Prentice-Hall Inc.

Parsons, Talcott. 1973. A functional theory of change. In Amitai Etzioni, Ed., *Social Change*. New York: Basic Books.

Patin, Etienne and Lluis Quintana-Murci. 2008. Demeter's legacy: rapid changes to our genome imposed by diet. *Trends in Ecology and Evolution* **23**:56–59.

Peel, J. D. Y. 1971. *Herbert Spencer: The Evolution of a Sociologist*. New York: Basic Books.

Pennebaker, James W. 1980. Perceptual and environmental determinants of coughing. *Basic and Applied Social Psychology* **1**:83–91.

Perry, G. H., Nathaniel J. Dominy, Katrina G. Claw *et al.* 2007. Diet and the evolution of human amylase gene copy number variation. *Nature Genetics* **39**:1256–1260.

Petroski, Henry. 1992. *The Evolution of Useful Things*. New York: Alfred A. Knopf, Inc.

Phillimore, Albert B. and Trevor D. Price. 2008. Density-dependent cladogenesis in birds. *PLOS Biology* **6**:483–489.

Phillips, David P. 1977. Motor-vehicle fatalities increase just after publicized suicide stories. *Science* **196**:1464–1465.

Phillips, David P. 1978. Airplane accident fatalities increase just after newspaper stories about murder and suicide. *Science* **201**:748–750.

Phillips, David P. 1979. Suicide, motor vehicle fatalities, and the mass media: Evidence toward a theory of suggestion. *American Journal of Sociology* **84**:1150–1174.

Phillips, David P. 1980. Airplane accidents, murder, and the mass media: Towards a theory of imitation and suggestion. *Social Forces* **58**:1001–1024.

Phillips, David P. 1982. The impact of fictional television stories on U.S. adult fatalities: New evidence on the effects of violence in the mass media. *American Journal of Sociology* **87**:1340–1359.

Phillips, David P. 1983. The impact of mass media violence on homicides. *American Sociological Review* **48**:560–568.

Phillips, David P. and L. Carstensen. 1986. Clustering of teenage suicides after television news stories about suicide. *New England Journal of Medicine* **315**:685–689.

Phillips, David P. and D.J. Paight. 1987. The impact of televised movies about suicide: A replicative study. *New England Journal of Medicine* **317**:809–811.

Pianka, Eric R. 1970. On r- and K- selection. *The American Naturalist* **104**:592–597.

Pigliucci, Massimo. 2001. *Phenotypic Plasticity: Beyond Nature and Nurture.* Baltimore: The Johns Hopkins University Press.

Pinch, Trevor. 1986. *Confronting Nature: The Sociology of Solar Neutrino Detection.* Dordrecht: D. Reidel Publishing Co.

Pinker, Steven. 1994. *The Language Instinct: The New Science of Language and Mind.* London: Allen Lane.

Pittendrigh, C. S. 1958. Adaptation, natural selection, and behavior. Pp. 390–416 in A. Roe and George G. Simpson, Eds., *Behavior and Evolution.* New Haven: Yale University Press.

Platnick, Norman I. and H. Don Cameron. 1977. Cladistic methods in textual, linguistic, and phylogenetic analysis. *Systematic Zoology* **26**:380–385.

Plotkin, Henry. 1994. *Darwin Machines and the Nature of Knowledge.* Cambridge: Harvard University Press.

Pocklington, Richard and Michael L. Best. 1997. Cultural evolution and the units of selection in replicating texts. *Journal of Theoretical Biology* **188**:79–87.

Popper, Karl R. 1957. *The Poverty of Historicism.* London: Routledge & Kegan Paul.

Posner, Richard A. 1972, 2007. *Economic Analysis of Law*, 1st–7th editions. Boston: Little Brown. Reprinted New York: Aspen Publishers.

Powell, Russell A., Diane G. Symbaluk and Suzanne E. Macdonald. 2005. *Introduction to Learning and Behavior*, 2nd edn. Australia: Thomson Wadsorth

Prather, J.F., S. Peters, S. Nowicki and R. Mooney. 2008. Precise auditory-vocal mirroring in neurons for learned vocal communication. *Nature* **451**:305–310.

Price, G. R. 1972. Extension of covariance selection mathematics. *Annals of Human Genetics* **35**:485–490.

Pringle, J. W. S. 1951. On the parallel between learning and evolution. *Behaviour* **3**:174–215.

Promislow, Daniel E.L. and Paul H. Harvey. 1990. Living fast and dying young: A comparative analysis of life history variation among mammals. *Journal of Zoology* **220**:417–437.

Prothero, Donald R. 2007. *Evolution: What the Fossils Say and Why it Matters.* New York: Columbia University Press.

Rachlin, Howard. 1991. *Introduction to Modern Behaviourism*, 3rd edn. New York: W. H. Freeman & Co.

Raff, Rudolf A. 1996. *The Shape of Life: Genes, Development, and the Evolution of Animal Form.* Chicago: The University of Chicago Press.

Read, Andrew F. and Paul H. Harvey. 1989. Life history differences among the eutherian radiations. *Journal of Zoology (London)* **219**:329–353.

Rendell, Luke and Hal Whitehead. 2001. Culture in whales and dolphins. *Behavioral and Brain Sciences* **24**:309–324.

Reznick, David N. and Robert E. Ricklefs. 2009. Darwin's bridge between micro-evolution and macroevolution. *Nature* **457**:837–842.

Rheinberger, Hans-Jörg J. and Steffan Muller-Willie. 2004. Gene. In *Stanford Encyclopedia of Philosophy.* http://plato.stanford.edu/

Richerson, Peter J. and Robert Boyd. 2005. *Not by Genes Alone: How Culture Transformed Human Evolution.* Chicago: The University of Chicago Press.

Ridley, Matt. 1996. *The Origin of Virtue: Human Instincts and the Evolution of Cooperation.* New York: Viking.

Ridley, Mark. 2001. *The Cooperative Gene: How Mendel's Demon Explains the Evolution of Complex Beings.* New York: The Free Press.

Rilley, J. R., U. Greggers, A. D. Smith, D. R. Reynolds and R. Menzel. 2005. The flight path of honeybees recruited by the waggle dance. *Nature* **435**:205–207.

Ritchie, Michael G. 2007. Sexual selection and speciation. *Annual Review of Ecology, Evolution, and Systematics* **38**:79–102.

Ritt, Nikolaus. 2004. *Selfish Sounds and Linguistic Evolution: A Darwinian Approach to Language Change.* Cambridge: Cambridge University Press.

Ritzer, George. 2000. *Modern Sociological Theory,* 5th edn. McGraw-Hill Higher Education.

Rizzolatti, Giancomo, Luciano Fadiga, Leonardo Fogassi and Vittorio Gallese. 1996. Premotor cortex and the recognition of motor actions. *Cognitive Brain Research* **3**:131–141.

Roff, Derek A. 2002. *Life History Evolution.* Sunderland, Mass.: Sinauer Associates Inc.

Rogers, Everett M. 2003. *Diffusion of Innovations,* 5th edn. New York: The Free Press.

Romanes, George John. 1884. *Animal Intelligence.* New York: D. Appleton and Company.

Romer, Paul. 2008. Economic growth. Pp. 128–131 in David R. Henderson, Ed., *The Concise Encyclopedia of Economics.* Indianapolis: Liberty Fund.

Roughgarden, Joan. 2004. *Evolution's Rainbow: Diversity, Gender and Sexuality in Nature and People.* Berkeley: University of California Press.

Rousseau, Jerome. 2006. *Rethinking Social Evolution: The Perspective from Middle-Range Societies.* McGill-Queen's University Press.

Ruckstuhl, Kathreen E. and Peter Neuhaus. 2005. *Sexual Segregation in Vertebrates: Ecology of the Two Sexes.* Cambridge: Cambridge University Press.

Rueffler, Claus, Tom J. M. Van Dooren, Olaf Leimar and Peter A. Abrams. 2006. Disruptive selection and then what? *Trends in Ecology and Evolution* **21**:238–245.

Ruhlen, Merritt. 1994a. *On the Origin of Languages: Studies in Linguistic Taxonomy.* Stanford, CA: Stanford University Press.

Ruhlen, Merritt. 1994b. *The Origin of Language: Tracing the Evolution of the Mother Tongue.* New York: Wiley.

Runciman, Walter. 2001. From nature to culture, from culture to society. *Proceedings of the British Academy* **110**:235–254.

Runciman, Walter. 2002. Heritable variation and competitive selection as the mechanism of sociocultural evolution. *Proceedings of the British Academy* **112**:9–25.

Runciman, Walter G. 2005a. Culture does evolve. *History and Theory* **44**:1–13.

Runciman, Walter G. 2005b. Rejoinder to Fracchia and Lewontin. *History and Theory* **44**:30–41.

Ruse, Michael. 1997. *Monad to Man: The Concept of Progress in Evolutionary Biology.* Cambridge: Harvard University Press.

Ruse, Michael. 1999. *Mystery of Mysteries: Is Evolution a Social Construction?* Cambridge: Harvard University Press.

Ruse, Michael. 2006. *Darwinism and its Discontents.* Cambridge: Cambridge University Press.

Russell, W. M. S. 1962. Evolutionary concepts in behavioral science IV. *General Systems* **7**:157–193.

Saioud, Saleema and Marion Blute. 2006. Some empirical trends in theory. *Perspectives. Washington: Theory Section, American Sociological Association* **28**(3):12–14. http://www.asatheory.org/

Sanderson, Stephen K. 2007. *Evolutionism and Its Critics: Deconstructing and Reconstructing an Evolutionary Interpretation of Human Society*. Boulder: Paradigm Publishers.

Sarkar, Sahotra. 2008. A note on frequency dependence and the levels/units of selection. *Biology and Philosophy* **23**:217–228.

Schilthuizen, Menno. 2001. *Frogs, Flies, and Dandelions*. Oxford: Oxford University Press.

Searle, John R. 1995. *The Construction of Social Reality*. New York: The Free Press.

Searls, David B. 2002. The language of genes. *Nature* **420**:211–217.

Seger, Jon and H. June Brockmann. 1987. What is bet-hedging? Pp. 182–211 in P. Harvey and L. Partridge, Eds., *Oxford Surveys in Evolutionary Biology IV*. Oxford: Oxford University Press.

Semple, Charles and Mike Steel. 2003. *Phylogenetics*. New York: Oxford University Press.

Shennan, Stephen. 2008. Evolution in archaeology. *Annual Review of Anthropology* **37**:75–91.

Shuster, Stephen M., and Michael J. Wade. 2003. *Mating Systems and Strategies*. Princeton: Princeton University Press.

Simpson, George Gaylord. 1961. *Principles of Animal Taxonomy*. New York: Columbia University Press.

Singer, Peter. 2000. *A Darwinian Left: Politics, Evolution and Cooperation*. New Haven: Yale University Press.

Singh, Jitendra V., Ed., 1990. *Organizational Evolution: New Directions*. Newbury Park, CA: Sage Publications Ltd.

Sismondo, Sergio. 1993. Some social constructions. *Social Studies of Science* **23**:515–53.

Sismondo, Sergio. 1996. *Science Without Myth: On Constructions, Reality, and Social Knowledge*. Albany: State University of New York Press.

Skinner, B. F. 1948a. *Walden Two*. New York: Macmillan.

Skinner, B. F. 1948b. 'Superstition' in the pigeon. *Journal of Experimental Psychology* **38**:168–172.

Skinner, Brian F. 1957. *Verbal Behavior*. Englewood Cliffs, N. J.: Prentice-Hall.

Skinner, B.F. 1966. The phylogeny and ontogeny of behavior. *Science* **153**:1205–1213.

Skyrms, Brian. 2004. *The Stag Hunt and the Evolution of Social Structure*. Cambridge: Cambridge University Press.

Slater, P. J. B. and S. A. Ince. 1979. Cultural evolution of chaffinch song. *Behaviour* **71**:146–166.

Slater, P. J. B., S. A. Ince, and P. W. Colgan. 1980. Chaffinch song types: Their frequencies in the population and distribution between the repertoires of different individuals. *Behaviour* **75**:207–218.

Smil, Vaclav. 2008. *Energy in Nature and Society: General Energetics of Complex Systems*. Cambridge, Mass.: MIT Press.

Smith, Adam. 1776, 1977. *An Enquiry Into the Nature and Causes of the Wealth of Nations*. Chicago: University of Chicago Press.

Smith, Robert L. 1984. *Sperm Competition and the Evolution of Animal Mating Systems*. Orlando: Academic Press, Inc.

Smolin, Lee. 1997. *The Life of the Cosmos*. Cambridge: Cambridge University Press.

Snopes.com. 2007. Horse's pass. http://snopes.com/history/american/gauge.htm

Sober, Eliot and David S. Wilson. 1998. *Unto Others: The Evolution and Psychology of Unselfish Behavior*. Cambridge, Mass.: Harvard University Press.

Solomon, Richard. 1998. Case closed? Re: origin of US standard railway gauge. http://lists.essential.org/1998/am-info/msg01555.html

Spector, Rosanne. 2005. Me-too drugs: Sometimes they're just the same old, same old. *Stanford Medicine Magazine* **22**(2) Summer.

Spencer, Herbert. 1862. *First Principles*. London: Williams & Norgate.

Spencer, Herbert. 1873. *The Study of Sociology*. London: Williams & Norgate.

Spencer, Herbert. 1904. *An Autobiography*. (2 volumes.) London: Williams and Norgate.

Staddon, John E. R. 1975. Learning as adaptation. Pp. 37–98 in Wm. K. Estes, Ed., *Handbook of Learning and Cognitive Processes*, Vol. II. Hillsdale, N. J.: Lawrence Erlbaum Associates Publishers.

Staddon, John E. R. and Virginia L. Simmelhag. 1971. The superstitious experiment: A reexamination of its implications for the principles of adaptive behaviour. *Psychological Review* **78**:3–43.

Stanovich, Keith E. 2004. *The Robot's Rebellion: Finding Meaning in the Age of Darwin*. Chicago: The University of Chicago Press.

Stearns, Stephen C. 1992. *The Evolution of Life Histories*. Oxford: Oxford University Press.

Stegmann, Ulrich E. 2004. The arbitrariness of the genetic code. *Biology and Philosophy* **19**:205–222.

Stone, J. R. 1996. The evolution of ideas: A phylogeny of shell models. *The American Naturalist* **148**:904–929.

Stotz, Karola, Paul E. Griffiths and Rob Knight. 2004. How biologists conceptualize genes: an empirical study. *Studies in History and Philosophy of Biological and Biomedical Sciences* **35**:647–673.

Sugiyama, Y. and J. Koman. 1979. Tool-using and -making behavior in wild chimpanzees at Boussou, Guinea. *Primates* **20**:513–524.

Sutherland, William J. 2003. Parallel extinction risk and global distribution of languages and species. *Nature* **423**:276–279.

Swanson, Carl P. 1983. *Ever-Expanding Horizons: The Dual Informational Sources of Human Evolution*. Amherst: The University of Massachusetts Press.

Swidler, Ann. 1986. Culture in action: Symbols and strategies. *American Sociological Review* **51**:273–286.

Tardé, Gabriel. 1903. *The Laws of Imitation*. Translated from the second French edition in 1962 by Elsie Clews Parsons with an introduction by Franklin H. Giddings. New York: Henry Holt & Company. Reprinted, 1962. Gloucester, Mass.: Peter Smith.

Tëmkin, Ilya and Niles Eldredge. 2007. Phylogenetics and material cultural evolution. *Current Anthropology* **48**:146–153.

Teresi, Dick. 2002. *Lost Discoveries: The Ancient Roots of Modern Science – From the Babylonians to the Mayas*. New York: Simon & Schuster.

Thorndike, Edward L. 1911, 2000. *Animal Intelligence: Experimental Studies*. New Brunswick, N. J.: Transaction Publishers.

Thorpe, W. H. 1958. The learning of song patterns by birds, with especial reference to the song of the chaffinch *Fringilla coelebs*. *Ibis* **100**:535–570.

Tilley, Christopher, Webb Keane, Susanne Küchler, Michael Rowlands, and Patricia Spyer, Eds., 2006. *Handbook of Material Culture*. London: Sage Publications.

Tilly, Charles. 2006. http://www.professor-murmann.info/tilly/2006_History of_in_Sociology.pdf

Tishkoff, Sarah A., Mary K. Gonder, Brenna M. Henn *et al.* 2007. History of click-speaking populations of Africa inferred from mtDNA and Y chromosome genetic variation. *Molecular Biology and Evolution* **24**:2180–2195.

Tolman, Edward Chance. 1949. *Purposive Behavior in Animals and Man*. Berkeley: University of California Press.

Tolman, Edward Chance, B. F. Ritchie and D. Kalish. 1946. Studies in spatial learning II Place learning versus response learning. *Journal of Experimental Psychology* **36**:221–226.

ToL. 1996–2009. *Tree of Life Web Project*. http://www.tolweb.org/tree/

Tomasello, Michael. 2008. *Origins of Human Communication*. Cambridge: The MIT Press.

Tottenham, Laurie Sykes, Deborah Saucier, Lorin Elias and Carl Gutwin. 2003. Female advantage for spatial location memory in both static and dynamic environments. *Brain and Cognition* **53**:381–383.

Trivers, Robert L. 1971. The evolution of reciprocal altruism. *Quarterly Review of Biology* **46**:35–57.

Trivers, Robert L. 1972. Parental investment and sexual selection. Pp. 136–179 in Bernard Campbell, Ed., *Sexual Selection and the Descent of Man, 1871–1971*. New Brunswick, N. J.: Aldine Transaction.

Turke, Paul E. 2008. William's theory of the evolution of senescence: Still useful at fifty. *The Quarterly Review of Biology* **83**:243–256.

Tversky, Amos and Daniel Kahneman. 1992. Advances in prospect theory: Cumulative presentation of uncertainty. *Journal of Risk and Uncertainty* **5**:297–323.

Tylor, Edward B. 1871, 1958. *Primitive Culture V I: Researches into the Development of Mythology*. London: J. Murray. Reprinted in Gloucester, Mass.: P. Smith.

Van Driem, George. 2001. *Languages of the Himalayas: An Ethnolinguistic Handbook of the Greater Himalayan Region Containing an Introduction to the Symbiotic Theory of Language*. Leiden; Boston:V. I. Brill.

Van Parijis, Philippe. 1981. *Evolutionary Explanations in the Social Sciences: An Emerging Paradigm*. London: Tavistook Publications.

Van Valen L. 1973. Fetschrift. *Science* **180**:488.

Van Wyhe, John. 2002–9. *The Complete Works of Charles Darwin Online*. http://darwin-online.org.uk

Vaughan, William and Richard J. Herrnstein. 1987. Stability, melioration and natural selection. Pp. 185–215 in L. Green and J. H. Kagel, Eds., *Advances in Behavioral Economics* I. Norwood, N. J.: Ablex.

Verdu, Paul, Frederic Austerlitz, Arnaud Estoup *et al.* 2009. Origins and genetic diversity of pygmy hunter-gatherers from western central Africa. *Current Biology* **19**:312–318.

Von Neumann, John and Oskar Morgenstern. 1944. *Theory of Games and Economic Behavior*. Princeton: Princeton University Press.

Walsh, Denis. 2008. Teleology. Pp. 113–137 in Michael Ruse, Ed., *The Oxford Handbook of Philosophy of Biology*. Oxford: Oxford University Press, Inc.

Watson, Richard A. 2008. *Compositional Evolution: The Impact of Sex, Symbiosis, and Modularity on the Gradualist Framework of Evolution*. Cambridge: MIT Press.

Weber, Max. 1904–5, 1958. *The Protestant Ethic and the Spirit of Capitalism*. Translated by Talcott Parsons. New York: Charles Scribner's Sons.

West, Geoffrey B., James H. Brown and Brian J. Enquist. 1997. A general model for the origin of allometric scaling laws in biology. *Science* **276**:122–126.

West, Stuart A., Ashleigh S. Griffin and Andy Gardner. 2007. Evolutionary explanations for cooperation. *Current Biology* **17**:R661-R672.

West-Eberhard, Mary Jane. 2003. *Developmental Plasticity and Evolution*. Oxford: Oxford University Press.

Wheeler, Michael, John Ziman and Margaret A. Boden, Eds., 2002. *The Evolution of Cultural Entities*. Oxford: Oxford University Press.

Whitehead, Hal, Luke Rendell, Richard W. Osborne, and Bernd Würsig. 2004. Culture and conservation of non-humans with reference to whales and dolphins: Review and new directions. *Biological Conservation* **120**:427–437.

Whitehouse, Harold L. K. 1965, 1969, 1973. *Towards an Understanding of the Mechanism of Heredity*. London: Edward Arnold.

Whiten, A., J. Goodall, W. C. McGrew *et al.* 1999. Culture in Chimpanzees. (With an introduction by Frans de Waal.) 1999. *Nature* **399**:682–685 and 635–636.

Whiten, A., J. Goodall, W. C. McGrew *et al.* 2001. Charting cultural variation in Chimpanzees. *Behaviour* **138**:1481–1516.

Whiten, Andrew, Victoria Horner, and Frans B. M. de Waal. 2005. Conformity to cultural norms of tool use in chimpanzees. *Nature* **437**:737–740.

Whiten, Andrew, Antoine Spiteri, Victoria Horner *et al.* 2007. Transmission of multiple traditions within and between Chimpanzee groups. *Current Biology* **17**:1038–1043.

Whitfield, John. 2004. Ecology's big, hot idea. *PloS Biology* **2**:2023–2027.

Whitfield, John. 2006. *In the Beat of a Heart: Life, Energy, and the Unity of Nature*. Washington, D.C.: Joseph Henry Press.

Wikipedia. 2009. *List of Cognitive Biases*. http://en.wikipedia.org/wiki/List_of_cognitive_biases

Williams, George C. 1966. *Adaptation and Natural Selection: A Critique of Some Current Evolutionary Thought*. Princeton: Princeton University Press.

Williams, George C. 1992. *Natural Selection: Domains, Levels, and Challenges*. Oxford: Oxford University Press.

Williams, George C. 1997. *The Pony Fish's Glow and Other Clues to Plan and Purpose in Nature*. New York: Basic Books.

Williamson, Oliver E. 1995. *Transaction Cost Economics*. U.K.: Edward Elgar Publishing.

Wilson, David Sloan. 2007. *Evolution for Everyone: How Darwin's Theory Can Change the Way We Think About Our Lives*. New York: Delacorte Press.

Wilson, David Sloan and Edward O. Wilson. 2007. Rethinking the theoretical foundations of sociobiology. *The Quarterly Review of Biology* **82**:327–348.

Wilson, Edward O. 1975. *Sociobiology: The New Synthesis*. Cambridge: Belknap Press of Harvard University Press.

Wimsatt, W. C. and C. Schank. 1988. Two constraints on the evolution of complex adaptations and the means for their avoidance. Pp. 231–273 in M. H. Nitecki, Ed., *Evolutionary Progress*. Chicago: University of Chicago Press.

Wittebolle, Lieven, Massimo Marzorati, Lieven Clement *et al.* 2009. Initial community evenness favours functionality under selective stress. *Nature* **458**:623–626.

Witting, Lars. 2008. Inevitable evolution: back to The Origin and beyond the 20th century paradigm of contingent evolution by historical natural selection. *Biological Reviews* **83**:259–294.

Wolfe, Jeremy M. 2006. *Sensation and Perception*. Sunderland, MA.: Sinauer Associates.

Wright, Robert. 2001. *Nonzero: The Logic of Human Destiny*. New York: Vintage Books.

Wrong, Dennis H. 1961. The oversocialized concept of man in modern sociology. *American Sociological Review* **26**:183–193.

Zentall, Thomas R. 2006. Imitation: definitions, evidence and mechanisms. *Animal Cognition* **9**:335–353.

Ziman, John. 1968. *Public Knowledge: An Essay Concerning the Social Dimension of Science*. Cambridge: Cambridge University Press.

Ziman, John, Ed., 2000. *Technological Innovation as an Evolutionary Process*. New York: Cambridge University Press.

Index

Printed in the United States
by Baker & Taylor Publisher Services